The Wave Function

The Wave Function

Essays on the Metaphysics of Quantum Mechanics

EDITED BY
ALYSSA NEY
AND
DAVID Z ALBERT

OXFORD
UNIVERSITY PRESS

OXFORD
UNIVERSITY PRESS

Oxford University Press is a department of the University of Oxford.
It furthers the University's objective of excellence in research, scholarship,
and education by publishing worldwide.

Oxford   New York
Auckland   Cape Town   Dar es Salaam   Hong Kong   Karachi
Kuala Lumpur   Madrid   Melbourne   Mexico City   Nairobi
New Delhi   Shanghai   Taipei   Toronto

With offices in
Argentina   Austria   Brazil   Chile   Czech Republic   France   Greece
Guatemala   Hungary   Italy   Japan   Poland   Portugal   Singapore
South Korea   Switzerland   Thailand   Turkey   Ukraine   Vietnam

Oxford is a registered trademark of Oxford University Press
in the UK and certain other countries.

Published in the United States of America by
Oxford University Press
198 Madison Avenue, New York, NY 10016

Library of Congress Cataloging-in-Publication Data
The wave function: essays on the metaphysics of quantum mechanics / edited by Alyssa Ney
and David Z Albert.
p. cm.
ISBN 978-0-19-979080-7 (hardcover: alk. paper)—ISBN 978-0-19-979054-8 (pbk.: alk. paper)
1. Quantum theory—Philosophy.   2. Metaphysics.   3. Wave functions.
I. Ney, Alyssa.   II. Albert, David Z   III. Title: Metaphysics of quantum mechanics.
530.1201—dc23   2012013512

ISBN 978-0-19-979080-7
ISBN 978-0-19-979054-8

Contents

Contributors

DAVID Z ALBERT, Columbia University

VALIA ALLORI, Northern Illinois University

STEVEN FRENCH, University of Leeds

SHELDON GOLDSTEIN, Rutgers University

PETER J. LEWIS, University of Miami

TIM MAUDLIN, New York University

BRADLEY MONTON, University of Colorado, Boulder

ALYSSA NEY, University of Rochester

JILL NORTH, Cornell University

DAVID WALLACE, University of Oxford

NINO ZANGHÌ, University of Genoa

Preface

DAVID Z ALBERT AND ALYSSA NEY

Wave functions, or some mathematical equivalent of wave functions, come up in every quantum theory and in every proposal for making explicit conceptual sense of the quantum theories that we presently have. We encounter them in nonrelativistic quantum mechanics, and in relativistic quantum field theories, and in supersymmetric quantum string theories, and in the Copenhagen interpretation of quantum mechanics, and in the many-worlds interpretation of quantum mechanics, and in Bohmian mechanics, and in theories of spontaneous state reduction, and they may well amount (in short) to an indispensable element of any scientific account of anything even remotely like the world we live in.

This book contains ten newly commissioned essays addressed to the question of what kinds of things wave functions are. The essays were written with an audience of philosophers—mainly philosophers of science and metaphysicians—in mind, and so they will not presuppose a background in physics. Where equations are presented, they are explained in a nontechnical way, and the book begins with an extensive introduction that describes the history of this debate and contains an elementary exposition of the physics with which the later chapters engage. We thus intend this book to be a self-contained introduction to the topic, and we hope these essays will be a launching pad for more work on the foundational issues they raise. But this book is also addressed to anyone who is interested in learning about how physics, particularly quantum theory, describes the world as it is at its most fundamental.

The past few decades have seen an upswing in work at the intersection of philosophy and physics in which writers are reengaging with questions the founders of quantum mechanics asked. Someone familiar merely with popular presentations of the science might be surprised what the central questions are. A common representation of the early philosophical struggles over quantum mechanics presents it as a debate between those like Bohr and Heisenberg on the one hand, who showed us that the universe is fundamentally indeterministic and only there when we are looking, and a caricature (more likely misrepresentation altogether) of an Einstein who dogmatically rejected indeterminism and the revolutionary rejection of objectivity the founders (Bohr and Heisenberg) saw. It isn't that this is all incorrect. Einstein did say, "God doesn't play dice" (or something very similar).[1] And Heisenberg certainly did insist that concepts entering physical theories must be limited to what may be observed. But the current debates—those represented in this volume—are not primarily taken up with these issues.

First, the essays presuppose a scientific realist attitude to fundamental physics. That is, these essays presuppose the view that our fundamental physical theories are engaged in the project of coming to understand the world as it is independent of our experience of it. This is not the view commonly attributed to Bohr and Heisenberg. Our point here is that the question is not whether quantum mechanics can describe *a* world as it is independent of us as observers. Rather, the question is how quantum mechanics describes *the* world as it is independent of us as observers. One of the central questions this volume addresses is whether quantum mechanics tells us that the world is fundamentally made of a wave function or wave functions.

Second, the issue of determinism versus indeterminism—whether the fundamental laws of the universe incorporate genuine chances—is not one of the main issues at play in the essays that follow.[2] Most of the authors herein are officially neutral about whether the correct version of quantum mechanics is an indeterministic or a deterministic theory. All will agree that one of the central laws of nonrelativistic quantum mechanics, the Schrödinger equation, is deterministic. All will agree that one has to at the same time acknowledge that the sort of evidence we have for quantum mechanics is not uniformly about what does happen as a matter of necessity but also much of the time about what proves likely to happen with such and such probability. We have support for the Born rule: that the probability of an event's occurrence is related to the square amplitude of its wave function. But whether these seeming facts about probabilities must be understood as deriving from what are ultimately indeterministic laws or have more to do with something about ourselves, the consequences of our actions or the limits of our knowledge is often left open.

Still, the debates about what quantum theory tells us about the nature of our world are descended from debates the founders of quantum mechanics were hav-

---

[1] To Max Born, Einstein wrote, "I, at any rate, am convinced that *He* is not playing at dice," in Irene Born's translation (Born and Einstein 1971, p. 88).

[2] Readers who are already familiar with the debates in this volume know that this is a subtle point. One of the main issues about whether Everettian versions of quantum mechanics can be correct has to do with whether in an Everettian universe there can be the sort of probabilities we think quantum mechanics teaches us about (and, if not, whether this is a problem). Note, though, that the trouble here isn't supposed to be that Everettian quantum mechanics is deterministic. Bohmians also think the fundamental laws are all deterministic, and they also make this objection to the Everettian. The trouble is supposed to be that there is no way to capture these facts about probabilities, either in the fundamental laws of Everettian quantum mechanics or in any derivative way.

ing in the 1920s and 1930s. As James Cushing (1994) has argued, one of the earliest debates between those like de Broglie, Einstein (perhaps), and Schrödinger, on one hand, and Bohr, Heisenberg, and Pauli, on the other, was about whether the quantum state (the wave function) could be a complete description of reality or whether quantum mechanics as it stood needed supplementation with additional variables to connect its claims to the world we experience. And as Arthur Fine (1986) has argued, Einstein's representation in these debates as a spoilsport who could not come to terms with the radical new ideas of the quantum theory is undermined by evidence from his own writings and letters. Indeed, Fine argues, it was Einstein who (with Schrödinger) was more receptive to a break with classical concepts like position and momentum if that is where the new ideas led, whereas Heisenberg and especially Bohr with his correspondence principle, insisted on retaining contact with the classical concepts.

One also finds in the early history of quantum mechanics one of the main issues taken on in some of the essays here about the nature of the spatial structure presupposed by the fundamental ontology of quantum mechanics. Early in the development of his theory, Schrödinger came to realize that taking his mathematical framework—particularly its central item of representation, the wave function—with ontological seriousness would require viewing the fundamental physical space not as the ordinary three-dimensional space of our acquaintance but as a higher-dimensional and extremely unfamiliar configuration space. He took this to be a serious problem that needed addressing.[3] Some of the essays here ask whether it may yet be a consequence we should take on board.

These are some of the questions with which the essays in this volume engage. We hope the discussions will be insightful for those metaphysicians wishing to think more about the fundamental ontology and spatial structure of our world. We also hope it will stimulate more discussion between metaphysicians and philosophers of physics on topics of fundamental metaphysics.

Finally, we hope this book will stimulate more discussions between philosophers and physicists. In philosophy, we sometimes hear physicists caricatured as nearly universally taking Feynman's "Shut up and calculate!" attitude to their own theories. Yet these problems in the metaphysics of physics are problems in the foundations of physics. They are problems that lie at the intersection of philosophy and physics. Lee Smolin's (2006) book *The Trouble with Physics* lists the five great outstanding problems in theoretical physics. The second is the following: "Resolve the problems in the foundations of quantum mechanics, either by making sense of the theory as it stands or by inventing a new theory that does make sense" (p. 8). As philosophers, our role isn't to invent new physical theories. But we do have something to contribute to solving this problem. We hope these essays will stimulate those who are already engaged in this joint project with physics and tempt more to join the debate.

---

[3] See the letters in Przibram (1967), especially those between Schrödinger and Lorentz.

We thank all of our contributors for taking the time to write these new essays. We also thank Peter Ohlin of Oxford University Press for his support and encouragement as we worked on this project, and all of the metaphysicians who offered generous feedback on how the work could best serve its targeted audience. Finally, we thank Barry Loewer for many valuable discussions on the topics discussed here, helping both of us as we've struggled to understand these metaphysical issues more deeply.

References

Born, Max, and Albert Einstein. 1971. *The Born-Einstein Letters 1916–1955*. London: Macmillan.

Cushing, James T. 1994. *Quantum Mechanics: Historical Contingency and the Copenhagen Hegemony*. Chicago: University of Chicago Press.

Fine, Arthur. 1986. *The Shaky Game: Einstein, Realism, and the Quantum Theory*. Chicago: University of Chicago Press.

Przibram, K. 1967. *Letters on Wave Mechanics*. New York: Philosophical Library.

Smolin, Lee. 2006. *The Trouble with Physics: The Rise of String Theory, the Fall of Science, and What Comes Next*. New York: Spin Network.

The Wave Function

Introduction[*]

ALYSSA NEY

The purpose of this introduction is to provide the reader who is unfamiliar with contemporary debates in the metaphysics of quantum mechanics some background in the central issues that arise in this volume. Those familiar with the history and mathematical structure of quantum mechanics and the measurement problem may want to skip ahead to the discussion of metaphysical issues beginning in section 5. For those wanting more background, a brief history of quantum mechanics is contained in section 1. My goal there is to sketch one clean, direct, accessible path from classical physics to quantum theory that serves to clarify the main interpretive issues that concern us in this volume. For a more thorough treatment of the history of quantum theory, the reader is directed to consult one of the many histories that have been written on the subject.[1] Section 2 introduces some of the technical concepts crucial to understanding nonrelativistic quantum mechanics and gives a brief account of the representation of states, relationship of states to measurements, and dynamics. Section 3 introduces the measurement problem, and section 4 discusses solutions to it. Section 5 provides an overview of the main metaphysical options and their motivation. Section 6 considers the extension of these issues to a relativistic setting.

1 Early History of Quantum Mechanics

One of several key steps in the development of quantum theory was Niels Bohr's 1913 hypothesis about the structure of the atom. Bohr's hypothesis allowed resolution of a puzzle that had plagued physics. Given a model of the atom as a system containing negatively charged electrons circling a positively charged nucleus, electromagnetism dictated that the electrons would be attracted to the nucleus and drift toward it, soon spiraling in and collapsing the atom. Of course,

---

[*] I thank David Albert, Valia Allori, Peter Lewis, and David Wallace for generously taking the time to read and provide comments on an earlier draft of this introduction.

[1] For a nontechnical but thorough and trustworthy recent history of quantum mechanics, Kumar (2008) is recommended.

we know that the electrons in an atom do not behave this way, and this is why atoms are stable and matter exists as we know it. As Bohr put it in his Nobel Prize in Physics lecture (for the prize awarded in 1922):

> On the basis of our picture of the constitution of the atom it is thus impossible, so long as we restrict ourselves to the ordinary mechanical laws, to account for the characteristic atomic stability which is required for an explanation of the properties of the elements. (Bohr 1964, p. 11)

He solved this problem by suggesting that the set of allowable orbits, and hence energy levels, in an atom was not continuous but discrete. This is referred to as the "quantization of energy levels" in the atom. For an electron to spiral into the nucleus due to electrical attraction, there would need to be intermediate positions for it to be drawn to on its way in. If there were not such intermediate positions, there would be no drifting.

Bohr's proposal may be summarized by the following two postulates:

1. An atomic system can exist only in certain discrete "stationary states" which define corresponding allowed energy levels.
2. Only transitions from one of these allowed energy levels (E_1) to another (E_2) takes place. These transitions are accompanied by the emission or absorption of light of frequency ν according to the rule: $E_2 - E_1 = h\nu$, where h is Planck's constant. (Kemble 1958, p. 374)

One fascinating thing about these postulates, which Bohr deployed to explain not only the paradox of the stable atom but also long-held results about atomic spectra, was that it revealed an analogy between the behavior of electrons in atoms and waves. For $E = h\nu$ was Planck's law, proposed in 1900 about the relationship between the energy of a unit of electromagnetic radiation and its frequency. Max Planck had argued that electromagnetic radiation could only come in fixed units (or quanta), multiples of his constant h. This was the same relationship that Albert Einstein used in his 1905 explanation of the photoelectric effect.[2] Bohr exploited the same relationship Planck and then Einstein used to describe radiation to describe the energy states of electrons (matter).

In his explanation of the photoelectric effect, Einstein revived the particle theory of light earlier endorsed by Isaac Newton. This theory had been rejected in

---

[2] In some cases, when light shines on a piece of metal, electrons are knocked out of the metal. This is the photoelectric effect. However, for the effect to occur, the impinging light must be of a certain frequency (ν). A surprising observation was that increasing the intensity of the light did not affect the energy of the electrons leaving the metal. On the other hand, increasing the frequency of the light did. Einstein argued that this was because light is ultimately made up of particles (now called photons). Increasing the intensity of light only increases the number of photons, but does not increase the energy of the beam of light. To successfully get the electrons to release from the metal, one needed to increase the energy of the photons, not their quantity. Einstein used Planck's $E = h\nu$ to show why frequency mattered: photons with increased frequency had increased energy as well.

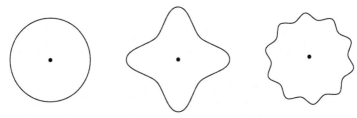

Figure I.1

favor of the wave theory of light since the early nineteenth century after Thomas Young showed the existence of interference phenomenon with his double-slit experiment with light. In his 1924 doctoral thesis, Louis de Broglie showed how Bohr's hypothesis about quantization in the atom could be explained by thinking of matter (in particular, electrons) as waves. Together these results led to widespread acceptance of the wave-particle duality of matter and radiation. Indeed, this duality was something that Bohr himself came to embrace as he developed his notion of complementarity—the view that microscopic phenomena had both wave and particle aspects, though these aspects are never revealed simultaneously and depend on the experimental context (Bohr 1927).

As we will see, de Broglie's work on Bohr's model of the atom reveals two different ways of interpreting the novel ideas of quantum theory. In Bohr's model, there was an emphasis on discontinuity: discontinuity in the discrete states that electrons orbiting a nucleus could occupy and discontinuity in the transitions that electrons were permitted between orbits. Because there were no allowable intermediate states, jumps between energy states had a peculiar mystery about them in Bohr's model. Particles seemed to jump instantaneously from here to a distant there—the quantum leap.

A virtue of de Broglie's thesis was that he was able to give an account of the quantization of energy states in the atom that lacked these aspects of discontinuity, exploiting a wave understanding of the electron. His idea was roughly the following (see Bohm 1951, p. 70 for more details). Start by thinking of the electron not as a particle circling the nucleus but as a wave wrapped around it. There are only so many ways for a wave to be wrapped around a nucleus completely. See figure I.1 for some examples.

In each case, what we find are integer multiples of complete wavelengths. Understanding electrons in this way as waves gives a motivation for the discreteness of energy levels. It also removed the need for discontinuity in the transitions between energy states.[3] So, it seemed, we could think of the electron as

---

[3] David Bohm describes the continuity in the de Broglie model in the following way: "If ... the electron can gain energy from some other system ... then we shall see that the wave gradually flows from its original toroid to another one corresponding to a higher energy level. While this process is taking place there is some probability that the particle can be found in either toroid. In fact, for neighboring energy levels, the toroids overlap to some extent so that the wave never goes from one region to another without crossing the intervening space, a necessary condition for continuity of flow" (1951, p. 76).

a particle and be faced with discontinuity, or we could think of the electron as a wave and have continuity. We return to this theme momentarily.

For some time after Bohr's initial proposal, there was a crisis. Physicists were becoming convinced by experimental results that the quantum theory was correct,[4] but did not yet possess a precise mathematical framework to represent physical states and transitions. Instead, quantum theory worked by using a messy assortment of new models combined with the use of what Bohr called the "correspondence principle," stating that any quantum predictions must match classical predictions in the macroscopic limit. Indeed, in this early period of quantum theory, the electrons themselves were still treated as classically behaving objects, only subject to the quantization of energy levels.

Fortunately, not one but two distinct mechanical frameworks were discovered for quantum theory in the 1920s. The first version, initially discovered by Werner Heisenberg and developed in collaboration with Max Born and Pascual Jordan in 1925, is matrix mechanics. This theory relied on new algebraic techniques (including the representation of states and observable properties by arrays of numbers (matrices)) that neatly captured the discontinuous states and transitions between energy levels contained in Bohr's model. The second version of quantum mechanics, wave mechanics, was discovered the next year by Erwin Schrödinger. Schrödinger was explicit that he was inspired to produce this account by the work of de Broglie. Wave mechanics was taken up immediately by the physics community because it had two virtues Heisenberg's account lacked. First, it relied on mathematics that was already familiar to most of the physics community (the solving of differential equations). Second, Schrödinger's wave mechanics was thought to provide better visualization of physical systems than did the matrix account, which insisted on a representation of particles and discontinuous transitions (see Born 1964, p. 261; Jammer 1966, pp. 271–72). By applying a familiar theoretical framework developed to model the behavior of waves, Schrödinger's theory allowed one to visualize quantum systems and their transitions.

It is well documented how Heisenberg resisted Schrödinger's wave version of quantum mechanics, even after Schrödinger proved that the two frameworks were mathematically equivalent.[5] For Heisenberg, one did not fully appreciate the lessons of quantum theory and the conceptual leaps it required unless one made discontinuity of states and transitions an essential feature of the mechanical framework. As his ideas developed, he came to insist that this discontinuity was the consequence of a more general view in which the only states that had reality were those that were observed (1930). Thus, the central objects in Heisenberg's framework were the observables represented by matrices. Still, the rest of the physics community gradually came to recognize the utility of both frameworks, especially as the mathematics central to the Heisenberg framework

[4] See Heisenberg (1930), chapter 1, for an account of the main experiments.
[5] See the discussion in Kumar (2008), especially chapter 9.

became more familiar. Even Bohr, who insisted with Heisenberg that reality should be limited to what is observed, used both wave mechanics and matrix mechanics in his work (see, for example, Bohr 1927, p. 74).

2 Quantum Mechanics: Key Concepts

In this section, we build up to an understanding of those key physical concepts one finds in the metaphysical discussions that follow. These discussions require some basic familiarity with the representation of physical states one finds in quantum mechanics, an account of the connection between physical states and measurement, as well as some central dynamical principles.

We have already seen that one of the major innovations in quantum theory was the proposal that states are quantized. As we will see, this gets captured in the options chosen for the mathematical representation of states. Two common ways that quantum mechanics represents physical states are (1) as vectors in Hilbert space and (2) as functions from elements of a configuration space to (complex) numbers. Let's focus on the Hilbert space representation first.

A vector is typically defined as a mathematical object that has two features: magnitude (or length) and direction.[6] We follow Paul Dirac and use the following bracket notation to talk about the vectors that are representing quantum states: |v>. In a more extensive treatment, one would see how vectors have very straightforward representations in terms of (single-columned) matrices, or arrays of numbers. I mention this only to give an idea of how the mathematics presented here relates to Heisenberg's matrix mechanics formulation of quantum mechanics. To keep this introduction relatively compact and because we are primarily trying to build toward an understanding of the wave function and its place in quantum mechanics, we won't be exploring the details of matrix representations of vectors in Hilbert space. Interested readers can find excellent exposition of the main ideas aimed at philosophers in Albert (1992), Hughes (1989), and Ismael (2009).

Hilbert space is a kind of vector space that, like all vector spaces, is closed under the operations of vector addition and scalar multiplication. In other words, if two vectors, |v> and |u> are elements of a Hilbert space, then so is their sum |v> + |u>; and if a vector |v> is an element of a Hilbert space, then so is any vector that results from multiplying |v> by a scalar. Vectors may be added by attaching one vector's tail to the tip of the other and connecting them to form a new vector (from the former's tip to the latter's tail), as in figure I.2.

When one multiplies a vector by a scalar a, the result is a new vector pointing in the same direction as the original vector but with a length a times as long.

---

[6] However, it is difficult to know exactly how literally to interpret this when the vectors, as in the case of quantum mechanics, take on complex values.

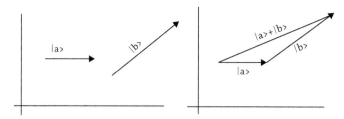

Figure I.2

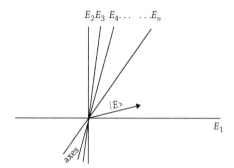

Figure I.3

One more feature of Hilbert space is that it is a vector space on which the notion of a dot product is defined. The dot product is an operation that multiplies two vectors and outputs a number. We can write the dot product of two vectors |A> and |B> as <A|B>. If <A|B> = 0, then these vectors are orthogonal to one another in the vector space.

In this approach, states are represented as normalized vectors in Hilbert space. All normalized vectors have length 1 (i.e., are unit vectors). Observable features are represented as operators on vectors, mathematical operators that act on vectors by modifying them in a particular way.[7]

The kind of Hilbert space used to represent a quantum state depends on the kind of system in which we are interested. Hilbert spaces used to represent states of position, momentum, or energy contain dimensions corresponding to allowable values of position, momentum, or energy that the particular system may have. We can represent a Hilbert space of the energy states of an electron graphically albeit incompletely as in figure I.3.

Each unit vector lying on an axis corresponds to a stationary state in the Bohr model. The dimensionality of such a Hilbert space will be infinite when there are an infinity of possible states: positions something could be in, possible momentum values, or energy states (even if such states are quantized).

---

[7] See Hughes (1989), chapter 2.

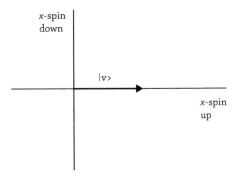

Figure I.4

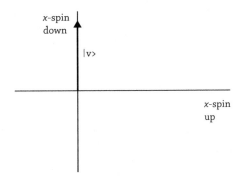

Figure I.5

We can also represent spin states using Hilbert space representations.[8] Particles like electrons may be described as having spin states in three dimensions (which we may label x, y, and z). To describe a particle's spin state, one focuses on the spin of a particle along a particular dimension, say, the x-dimension (x-spin). There are two determinate spin states for the electron in the x-dimension: x-spin up and x-spin down.[9] Because there are only two determinate x-spin states, the Hilbert space we use is a two-dimensional vector space. The vector used to represent an electron having x-spin up may then be written as shown in Figure I.4.

The vector used to represent an electron having x-spin down may then be written as shown in figure I.5.

Note that there are other possible vectors that can be seen to inhabit this same vector space of spin states. For example, the vector in figure I.6.

[8] Spin is commonly understood as an intrinsic angular momentum possessed by particles. Spin states were introduced to physics in 1925 by Samuel Goudsmit and George Uhlenbeck as quantum mechanics was being developed. Their work was a critical supplement to Bohr's model, required to give a more accurate model of atomic spectra.

[9] For other kinds of particles, there might be more possibilities.

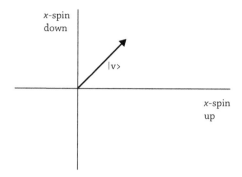

Figure I.6

Unlike those in figures I.4 and I.5, the vector in figure I.6 does not represent a particle in a determinate state of having x-spin up or x-spin down. Rather, this vector is pointing halfway between the previous two vectors. Using the law of vector addition illustrated in figure I.2, we can actually see that this vector can be written as a sum of two other vectors.

$$| \text{v} \rangle = \frac{1}{\sqrt{2}} |x\text{-spin up} \rangle + \frac{1}{\sqrt{2}} |x\text{-spin down} \rangle. \tag{I.1}$$

This is the first example we will see of a state of superposition. Vector |v⟩ in this example also represents a state of x-spin. It is a vector of length 1 (a normalized vector) in the two-dimensional Hilbert space we are using to represent x-spin states. However, this is not a representation of a state of an electron that determinately has x-spin up or x-spin down. This vector, we may say, has some of its magnitude pointing in the x-spin up direction and some of its magnitude pointing in the x-spin down direction. This is the representation of a state of an electron that is in a superposition of x-spin states. In Schrödinger's wave mechanics, the existence of such states (here, superpositions of x-spin; in the examples that follow, of position) follows from a classical wave mechanical principle: the principle of superposition. Much of what is at stake in the interpretation of quantum mechanics concerns how to understand states like this. We return to this issue of interpretation later.

Let's return to the representation of position states (or other coordinate space features like momentum or energy). Because there are often an infinite number of potential determinate position states when we are discussing a particle's location in a space that is continuous, the Hilbert spaces that are used to treat these kinds of states have an infinite number of dimensions. We can use a vector to represent a particle in a determinate position state, say, one located at (3, 0, 0). It is difficult to graphically represent such a thing, but here again, we may recognize that there are more vectors in this Hilbert space than just those used to represent determinate states of position. For example, consider a vector representing a state that is a superposition of the particle being located at (3, 0, 0) and being located at (4, 0, 0):

$$|v\rangle = \frac{1}{\sqrt{2}}\ |3, 0, 0\rangle + \frac{1}{\sqrt{2}}\ |4, 0, 0\rangle. \tag{I.2}$$

This vector has part of its magnitude in the (3, 0, 0) direction and part in the (4, 0, 0) direction. We can also discuss much more complicated superpositions, where the vector makes minor contributions to many of the infinite directions in the Hilbert space. If there were a finite number of available positions in ordinary space, any position state could be written down as the sum over the basic unit vectors that sit along the axes in the Hilbert space:

$$|v\rangle = a_1|v_1\rangle + a_2|v_2\rangle + a_3|v_3\rangle + \dots a_n|v_n\rangle \tag{I.3}$$

So that the resulting vector is also normalized, it would only be necessary that $a_1{}^2 + a_2{}^2 + a_3{}^2 + \dots + a_n{}^2 = 1$. As it happens, ordinary space is continuous, so position states cannot in general be written out as in (I. 3). Instead, a more general form of representation, in terms of functions, will be described in a moment.

Going back to the case of spin states, any spin state can be written down as the sum over the basic unit vectors that sit along the axes in that Hilbert space:

$$|v\rangle = a_1|x\text{-spin up}\rangle + a_2|x\text{-spin down}\rangle. \tag{I.4}$$

In general, when one specifies a set of unit vectors that are used to define the dimensions of such a Hilbert space, in terms of which a state may be represented as a sum, one is specifying what is called the "basis" of a vector space.[10]

It is worth noting for some of what follows that any Hilbert space can be coordinatized in a variety of ways, and thus the basis vectors that are chosen and determine the axes are to a certain extent arbitrary. It is probably easiest to illustrate this using the example of spin states.

I have said that we may talk about the spin states of particles along various dimensions in ordinary space, for example, *x*-spin, *y*-spin, *z*-spin. Experiments to detect the spin states of particles show that these sorts of features bear interesting relationships to one another. For example (and this fact is used when we discuss the measurement problem in section 3), when a particle has a determinate *z*-spin, say it has *z*-spin up, it will be in a superposition of *x*-spin states (see figure I.7) such that:

$$|z\text{-spin up}\rangle = \frac{1}{\sqrt{2}}\ |x\text{-spin up}\rangle + \frac{1}{\sqrt{2}}\ |x\text{-spin down}\rangle, \tag{I.5}$$

---

[10] Another way of understanding a basis is that for an N-dimensional Hilbert space, any collection of N mutually orthogonal vectors in that space forms a basis. An orthonormal basis is any collection of N mutually orthogonal vectors of length 1.

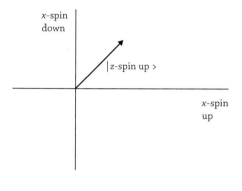

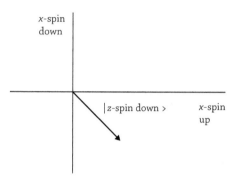

Figure I.7

and

$$|z\text{-spin down}\rangle = \frac{1}{\sqrt{2}}\,|x\text{-spin up}\rangle - \frac{1}{\sqrt{2}}\,|x\text{-spin down}\rangle. \qquad (I.\,6)$$

Note that these two new state vectors are also of length 1 and orthogonal to each other, so they form another orthonormal basis for spin space. We could instead, if we like, write down all x-spin states in terms of this z-spin basis.

What goes for spin states also goes for position, momentum, and energy states. There is always an infinite number of possible bases that can be used to write down states. Indeed, the famous EPR paper we discuss later exploits the interchangeability between position and momentum representations of one quantum state.

Sums of vectors are one way that states get represented in quantum mechanics. Another is using the concept of a wave function ψ, a function from elements in another kind of space, a configuration space,[11] to numbers. Often in textbooks

---

[11] For a single particle, configuration space is just ordinary space. More will be said below about configuration space for complex systems.

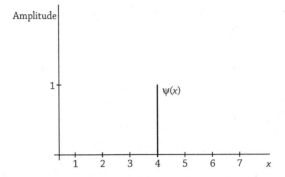

Figure I.8

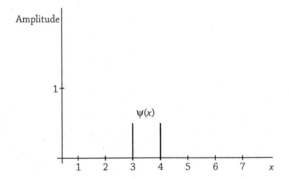

Figure I.9

the wave function is described as just another way to specify the state vector. Start by thinking about the position state of one point particle confined to a one-dimensional space with infinite length. We can represent this particle's position state as a function that takes as its input (domain) a point in that space and has as its output (range) a number. A requirement we put on these functions is that the sum of the squares of the numbers in the range must all add up to 1. It is natural to think of these numbers as amplitudes of the wave function at various points in configuration space, where in this case the configuration space is a one-dimensional line. This gives us a normalized wave function. For example, see figure I.8.

Figure I.8 represents the wave function of a particle that is determinately located at $x = 4$ in our one-dimensional space. Figure I.9 represents the wave function of a particle that is in a superposition of being located at $x = 3$ and $x = 4$.

Finally, figure I.10 represents the wave function of a particle that has high amplitude in the region around $x = 4$ and then dips off to have lower amplitude in other regions of space.

Given what we've already noted about the uncountable nature of the set of possible position states, an algebraic representation of states as a function describing the relevant curve over the space, in this case a Gaussian function, is

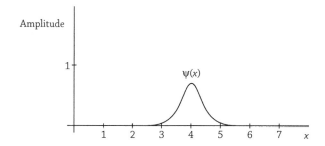

Figure I.10

more adequate than representation in terms of a sum of vectors. States like that in figure I.10 are typically thought to be more the norm in quantum mechanics rather than those in figures I.8 and I.9.

In a moment, we will extend this discussion to the representation of complex systems, that is, ones we can understand as comprising multiple particles. This will allow us to say more to clarify the nature of the configuration space used in the wave function representation of quantum mechanics (e.g., why it is called "configuration space"). But first let us say a bit more about quantum states in general and their connection with experiment. Given a system's state vector or wave function, one can infer the probabilities associated with the outcomes of various measurements one can make on the system. For example, for a state of the form:

$$\psi = a \, |A\rangle + b \, | \, B\rangle, \tag{I.7}$$

written down in a particular basis and normalized so that $a^2 + b^2 = 1$, if one makes a measurement on the system to find out whether it is in state $|A\rangle$ or state $|B\rangle$, the probability of finding the system in state $|A\rangle$ will be a^2. The probability of finding the system in state $|B\rangle$ will be b^2. For example, consider again our superposition of x-spin states:

$$\psi = \frac{1}{\sqrt{2}} \, |x\text{-spin up}\rangle + \frac{1}{\sqrt{2}} \, |x\text{-spin down}\rangle. \tag{I.8}$$

If we have an electron starting out in this state, if we were to measure its x-spin, we could predict that there is a $(\frac{1}{\sqrt{2}})^2$ chance, that is, a chance of 50%, that the electron will be found to have x-spin up and a $(\frac{1}{\sqrt{2}})^2$ chance, that is, a chance of 50%, that the electron will be found to have x-spin down.

In general, for a system represented by a wave function ψ, if one takes the absolute value of the wave function squared $|\psi|^2$ and integrates over a certain region of configuration space, this allows one to arrive at the probability of finding the system in a particular region of the space. This fact about the relation

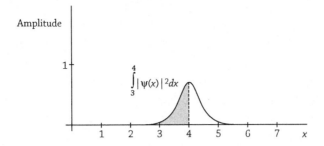

Figure I.11

between the wave function and the probabilities associated with measurement results is known as the Born rule. We can illustrate again using the case above, as in figure I.11.

Recall again that we are discussing a single-particle system in a one-dimensional space. Here the state is a function of position in that space and so can be written $\psi(x)$. In the diagram above, by calculating the area under the curve, we find the probability of finding the particle between $x = 3$ and $x = 4$. If, say, we want to know the probability of finding the particle somewhere in the region between $x = 2$ and $x = 5$, all we will do is integrate $|\psi|^2$ from $x = 5$ to $x = 2$:

$$\int_2^5 |\psi(x)|^2 \, dx. \qquad (I.9)$$

Using this method, we can infer probabilities for observing the electron at all regions in the space. If we integrate over the interval from $-\infty$ to $+\infty$, the result will be 1.

Now we can say a few things related to the interpretation of quantum states (i.e., wave functions) that come up naturally here. First, it is good to note that one important interpretation of the wave function is that it *is* just a summary of the probability of certain outcomes. This interpretation is usually associated with Born: interpret the wave function, or more precisely $|\psi|^2$, as a probability density for particles (Born 1964, p. 262). In other words, the wave function gives information about where it is more or less likely that one will find the particle, if one takes a measurement of its position.

This approach to understanding the wave function goes hand in hand with Heisenberg's way (which today we would call "antirealist") of understanding the lessons of quantum mechanics. Heisenberg is very explicit about this in the paper in which he offered his famous uncertainty principle, so it is worth a brief digression. The uncertainty principle states that there are certain pairs of observable quantities such that the more precisely one can know the value of one of the quantities, the less precisely one can know the value of the other. One such pair of quantities is position and momentum. Another is energy and time (in which the energy is being measured). Pairs of spin quantities are a third example, such as x-spin and z-spin. Heisenberg's argument for the uncertainty principle in the case of position and momentum is quite straightforward. Here is an excerpt:

> There is no shortage of such experiments, which in principle ... allow one to determine the "position of the electron" with arbitrary accuracy. For example, let one illuminate the electron and observe it under a microscope. Then the highest attainable accuracy in the measurement of position is governed by the wavelength of the light.... Every observation of scattered light coming from the electron ... can ... be so interpreted that a light quantum hits the electron, is reflected or scattered, and then, once again bent by the lens of the microscope, produces the photoeffect [i.e., the light that we observe]. At the instant when position is determined—therefore, at the moment when the photon is scattered by the electron—the electron undergoes a discontinuous change in momentum. This change is the greater the smaller the wavelength of the light employed—that is, the more exact the determination of the position. At the instant at which the position of the electron is known, its momentum therefore can be known up to magnitudes which correspond to that discontinuous change. (1927, p. 64)

Measuring the position of the electron affects its momentum. And, Heisenberg continues, measuring its momentum affects its position. The more precisely we are able to know the one, the less precisely we are able to know the other. From this epistemological point Heisenberg draws a metaphysical conclusion—that as we are able to narrow down a particle's position, it fails to have determinate momentum. He goes further, offering that in general what we are only ever able to know is the probability distribution of particles vis-à-vis observables like momentum and position: "In the fact that in quantum theory only the probability distribution of the position of the electrons can be given for a definite state ... one can recognize, with Born and Jordan, a characteristically statistical feature of quantum theory as contrasted to classical theory" (1927, p. 66). All measuring devices have a limit to the precision of which they are able measure. What cannot be measured does not exist.

Also in 1927, Bohr presented his principle of complementarity. Bohr's view was that certain pairs of features—particle/wave, position/momentum—are complementary, meaning they are never properly assigned to the same objects in the same circumstances. You could do an experiment to determine a particle's position, but in such a case momentum attributions are inappropriate and vice versa. He ties this principle directly to Heisenberg's claim. It is not possible to consider particles having the two features at once, because it is not possible to observe both features at once. In this way, the uncertainty principle is connected with certain antirealist ideas of Heisenberg and Bohr and the statistical inter-pretation of the wave function.

I note here that not all of the founders of quantum mechanics shared this reading of the wave function. In a letter to Born in 1927, Einstein refers to an attempt of his own to replace this interpretation of the wave function with a nonstatistical one.[12] Einstein's later remarks and papers reveal that in the end he decided on a distinct statistical understanding of the wave function in which it referred not to a description of a single system (say, the electron in question above) but an ensemble of similar systems.[13] The idea appears to be that the wave function describes what the position distribution would be for an ensemble (or collection) of systems similar to this one electron. Attempts to understand the wave function at this time thus seem to involve two options: (1) understand it in terms of measurements that we might make, or (2) understand it in terms of an imaginary ensemble of systems. As we will see, an alternative for the realist is to take the wave function at face value, as describing the state of a quantum sys-tem at a time. This is an approach that Schrödinger pursued early on (Przibam 1967). Although the quantum state may be connected with what we may expect of future measurements, this connection isn't analytic because the quantum state has a reality independently of measurements we might make of it.

Let us now extend our discussion to the representation of states of more than one particle. When we were focusing on the position state of one particle, we used a function from points in space to numbers. In a description of an ordinary particle allowed to move in three dimensions, this was a function from points of configuration space—in this case, ordinary, three-dimensional space to num-bers, the numbers we are calling the amplitudes of the wave function. The wave function can then be thought of as a field on three-dimensional space. When we have a system of two particles we want to describe, the wave function will again be a function from points in a configuration space to numbers. But now, the space in question is no longer three-dimensional. Instead, what is used is a six-dimensional configuration space.

Configuration space representation is not a new feature of quantum mechan-ics; it occurs in classical mechanics as well. Configuration space representations are useful ways of condensing information about the locations of many par-ticles in the following way. Where a system of N particles may be represented graphically as a system of N points in a three-dimensional coordinate space, it may alternatively be represented as one point in a three-times-N-dimensional

---

[12] In the letter he writes: "Last week I handed in a short paper to the Academy, in which I show that one can attribute quite definite movements to Schrödinger's wave mechanics, without any statistical interpretation." However, as Born notes in this collection of letters, this paper seems to have disap-peared (Born and Einstein 1971, p. 93).

[13] For discussion of Einstein's statistical interpretation, see Fine (1986), chapter 4.

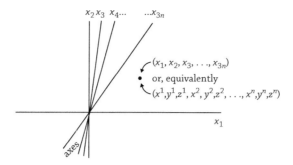

Figure I.12

configuration space. In general, the configuration space has three-times-N dimensions, where N is the number of particles in the system. Each point in the space can then be specified by an ordered $3N$-tuple, where the first three numbers correspond to the x, y, and z coordinates of the first particle, the next three numbers correspond to the x, y, and z coordinates of the second particle, and so on (see figure I.12).

The wave function can then be thought of as a function on this configuration space. Some, as we will see shortly, take the ontological step to thinking of it as a field on configuration space. The wave function takes (complex) numerical values at each point in configuration space. Just as a single particle system will be ordinarily thought of as spread out over space as in figure I.10, so multiparticle systems are typically thought of as spread out over higher-dimensional configuration space.

That the space on which the wave function is defined in quantum mechanics is a $3N$-dimensional configuration space, as opposed to three-dimensional ordinary space, was something that troubled Schrödinger as he tried to understand the wave function realistically. Part of the advantage of his wave mechanical approach was, we saw, that it provided a visualizable description of quantum systems. In the case of a one-particle system and its position representation, the particle could be visualized as a field—because its wave function was defined at every point in ordinary space with an amplitude at each point. When describing multi-particle systems, we may still interpret the wave function as a field, not on three-dimensional space but on $3N$-dimensional configuration space. Because this is not a space with which we are familiar, Schrödinger worried that this threatened the visualizability of the framework. However, as we will see, there is reason to think that this configuration space representation is an indispensable feature of quantum mechanics, even if philosophers may disagree about the ontological import of this means of representing states. This is another way in which quantum mechanics is so different from classical mechanics: there, configuration space representations are useful as they allow us very simple representations of complex systems. In quantum mechanics, they are more than useful if we are to give a complete characterization of the position states into which systems may enter—they appear to be required. Nonetheless, we can at least say that even in 3N-dimensional configuration space, the wave representation still has the virtue of giving us a continuous model of systems and transitions.

To see the importance of the 3N-dimensional representation, we must note that most quantum states of multiparticle systems have an interesting feature. They are what Schrödinger (1935, p. 161) called "entangled" states. He described such states in the following way:

> Maximal knowledge of a total system does not necessarily include total knowledge of all its parts, not even when these are fully separated from each other and at the moment are not influencing each other at all. Thus it may be that some part of what one knows may pertain to relations ... between the two subsystems (we shall limit ourselves to two), as follows: if a particular measurement on the first system yields *this* result, then for a particular measurement on the second the valid expectation statistics are such and such; but if the measurement in question on the first system should have *that* result, then some other expectation holds for that one the second.... In this way, any measurement process at all or, what amounts to the same, any variable at all of the second system can be tied to the not-yet-known value of any variable at all of the first, and of course *vice versa* also. (1935, p. 160)

Schrödinger then claims when there is entanglement, this "can obviously only go back to the fact that the two bodies at some earlier time formed in a true sense *one* system, that is were interacting, and have left behind *traces* on each other" (1935, p. 161). So entanglement arises as a matter of interaction between systems. What results is a complex state of the whole system that contains more information than can be inferred from the individual parts.

Here is a simplified example of an entangled state. Two particles interact at some time in the past and then have a wave function described by the following:

$$\psi = \frac{1}{\sqrt{2}} \, |3, 0, 0,>_1|7, 0, 0>_2 + \frac{1}{\sqrt{2}} \, |7, 0, 0>_1|3, 0, 0>_2. \qquad (I.10)$$

First, some notes on how to read this state. The subscripts next to the brackets indicate which particle is being described. What we are considering is a system of two particles in which the positions of the particles are tied to each other in the way Schrödinger was considering. The distinctive feature of an entangled state is that it cannot be written as the product of states of the individual systems. The probabilities of measurement results on either particle are tied to the other in a definite way. In fact, we can say that if we find the first particle at location $x = 3$, then with certainty we will find the second particle at $x = 7$. We know with certainty that if we find the second particle at $x = 3$, then with certainty we will find the first particle at $x = 7$. But we do not know the individual states of the particles, that is, where they are. We know they are separated by a distance of four units: that is, we know their relation to each other. But we do not have "total knowledge of the parts."

What is there to say about this sort of situation in metaphysical terms? We may start with the conjecture that the states of quantum mechanics are states of objects in ordinary three-dimensional space, but that when systems are in states like this, what we have are not really full fledged particles at the positions (3, 0, 0) and (7, 0, 0) but particle densities at the two locations, or in some sense parts of particles at these two points.[14] So what we have fundamentally is really the following: at location (3, 0, 0), a 50% part of electron 1 and a 50% part of electron 2, and at location (7, 0, 0), we have a 50% part of electron 1 and a 50% part of electron 1. However, this isn't sufficient information to capture the nature of the quantum state above.[15] For this proposal is consistent just as much with the state in equation (I.10) as with the following alternative state:

$$\psi_{1,2} = \frac{1}{\sqrt{2}}\ |3, 0, 0\rangle_1 |3, 0, 0\rangle_2 + \frac{1}{\sqrt{2}}\ |7, 0, 0\rangle_1 |7, 0, 0\rangle_2. \qquad (I.11)$$

However, the two quantum states described by equations (I.10) and (I.11) are measurably distinct. In one case, we can be sure to find the two electrons a distance of four units away from each other. In the other case, we can be sure to find them at the same location. The difference between these two states is captured, however, if we see them as fields in configuration space. To represent the first state, we have a field with amplitude 0.5 at location (3, 0, 0, 7, 0, 0) and amplitude 0.5 at location (7, 0, 0, 3, 0, 0) in configuration space. To represent the second state, we have a field with amplitude 0.5 at location (3, 0, 0, 3, 0, 0) and amplitude 0.5 at location (7, 0, 0, 7, 0, 0). To capture the difference between these two states, we must use the additional spatial degrees of freedom we have in configuration space. Entanglement is a completely pervasive feature of a quantum world, and what is needed to capture states that are entanglements of position like this is a configuration space representation.[16]

Schrödinger deployed the concept of entanglement in his famous cat example, and the same year, Einstein (with Boris Podolsky and Nathan Rosen) used it in their famous critique of the assumption that quantum mechanics provided a complete description of the world. In Schrödinger's scenario, states of a cat

---

[14] This is something like what Schrödinger tried to do as he was struggling with the nature of the configuration space representation of the wave function in 1926. See Przibram 1967, pp. 55–56.

[15] See Lewis 2004 for more details.

[16] It is worth mentioning that although I have focused so far on entanglements of position, entanglement is a much more general feature of quantum mechanics and can be seen to arise for cases of other features; see the case of spin states below.

(whether it is living or dead) are entangled with the states of a radioactive substance that has a 50% chance of decaying in a given hour. In the example, there is an opaque box that contains the cat and a flask containing poisonous gas. This flask is part of a "diabolical device": a hammer is set to break the flask if the radioactive substance decays, releasing the gas and killing the cat. After the hour has passed, the state of the system within the box may be represented as:

$$\psi = \frac{1}{\sqrt{2}} \; |\text{no gas released}\rangle_{\text{device}} |\text{alive}\rangle_{\text{cat}} + \frac{1}{\sqrt{2}} \; |\text{gas released}\rangle_{\text{device}} |\text{dead}\rangle_{\text{cat}}. \quad (\text{I}.12)$$

The state of the device is entangled with the state of the cat. Schrödinger uses this example to raise interpretive issues about the nature of the quantum state. Is the state of the whole device/cat system inside the box quite determinate and it is only our knowledge of what it is like that is fuzzy so we can only assign a 50% probability to the cat being alive? Or is it, rather, that the quantum state reveals everything that is true of the system at this time—so that as he says, "the living and the dead cat [are] (pardon the expression) mixed or smeared out in equal parts" (1935, p. 157)? As he notes in a lucid critique of the anti-realism of Bohr and Heisenberg, there is a difference between a blurry photograph and a photograph of clouds and fog banks. We return to the issue raised by Schrödinger's cat example—about the correct interpretation of the quantum state—in section 3.

The famous Einstein, Podolsky, and Rosen (EPR) article uses a microscopic example to illustrate a similar worry raised by the phenomenon of entanglement. This also concerns whether the wave function gives us an ontologically complete representation of a system at a time. We use the more familiar version of the EPR setup created by Bohm to describe the scenario. Imagine two electrons created in the entangled state of x-spin known as the singlet state:

$$\psi_{\text{singlet}} = \frac{1}{\sqrt{2}} \; |x\text{-spin up}\rangle_1 |x\text{-spin down}\rangle_2 - \frac{1}{\sqrt{2}} \; |x\text{-spin down}\rangle_1 |x\text{-spin up}\rangle_2. \quad (\text{I}.13)$$

Given such a state, we don't know what the individual x-spins of the electrons are. We only know that if we were to measure their individual x-spins, we would find that they were opposite of each other. If electron 1 has x-spin up, then electron 2 will have x-spin down, and vice versa. Now imagine separating the electrons so that they are at a great distance from one another. In fact, separate them so far that for a light signal from electron 1 to reach electron 2, it would take an hour. At noon, we measure the x-spin of electron 1 and find it to have x-spin up. What will happen if, five minutes later, someone else measures the x-spin of electron 2 where it is? Well, because the electrons were in the singlet state, we can know with certainty that electron 2 will have x-spin down. But how does this synchronization between the two electrons work? Before electron 1 was measured, there was a 50% chance that electron 2 would have x-spin up. If electron 1 hadn't been measured at noon, then there would be a 50% chance that electron 2 would have been found to have x-spin up at 12:05. But somehow, even though there is no time for a signal from electron 1 to reach electron 2 in time to "let it know" it had been observed x-spin up, electron 1's measurement changes the probability of electron 2 being measured as having x-spin up at 12:05 from 50% to 0%.

This is an example of the phenomenon of quantum nonlocality. Influences between quantum systems, in cases like this, seem to be carried instantaneously or at least superluminally. EPR suggested that perhaps cases like this showed that to avoid such nonlocality, we should deny that the quantum state is complete. They considered the possibility that perhaps there were further facts about the electrons that weren't captured by the quantum state. So in this case, the suggestion is that there is more to know about the spins of the electrons than is given by equation (I.13). Perhaps the wave function just describes what we can know about the system at a time, but there are further hidden facts about their individual spins that ensure their states are correlated in the appropriate way. In 1964, physicist John Bell proposed an experiment that (when later conducted successfully) would show that this speculation was wrong. He argued that there were setups that could be used to demonstrate the phenomenon of nonlocality where any prior synchronization of the two particles was impossible. Even if one presupposed there were hidden variables, facts about the spin states of the particles outrunning the quantum state described in equation (I.13), these facts wouldn't be sufficient to ground the connection between measurement results carried out on the pair. When Bell's predictions were confirmed, this established that EPR's suggestion that the quantum state is incomplete would not suffice to eliminate quantum nonlocality.[17]

We have now discussed representation of various kinds of quantum states and their connection with measurement. We now need to discuss the dynamics of quantum mechanics. For nonrelativistic systems, the law specifying how the state vector or wave function of any system changes over time is the time-dependent Schrödinger equation.[18] There are two features of the Schrödinger equation that are important to mention here. First, the Schrödinger equation is deterministic. That is, given the state of a system at a time, using this law, one can predict with certainty what the state of the system will be at some future time. Second, the law is linear, which entails the following:

---

[17] For accessible descriptions of Bell's argument addressed to philosophers, see Mermin (1981) and Maudlin (1994, chapter 1).

[18] The presentation of the dynamics here uses the Schrödinger equation because this is what will come up in the essays that follow this introduction. We have already noted that there is a mathematically equivalent alternative to Schrödinger's wave mechanical formulation of quantum mechanics—Heisenberg's matrix mechanics. This formulation makes use of a distinct equation, known as the Heisenberg equation. The central difference between the two equations is that whereas Schrödinger's equation is stated in terms of the evolution of a wave function over time, the Heisenberg equation focuses on a change in the observable features that a system has over time.

If |A> evolves to |A'> given the laws, and |B> evolves to |B'> given the laws, then a|A> + b|B> evolves to a|A'> + b|B'> given the laws.

This fact of linearity has an interesting consequence that will be relevant in the next section. To see this, consider the case of Schrödinger's cat already described. We wrote down the state of the system using the following wave function (equation I.12):

$$\psi = \frac{1}{\sqrt{2}} \text{ |no gas released>}_{\text{device}}\text{|alive>}_{\text{cat}} + \frac{1}{\sqrt{2}} \text{ | gas released>}_{\text{device}}\text{| dead>}_{\text{cat}}.$$

If this is the state of the system after an hour, then if the dynamical laws are linear, if the system obeys these laws, this state will never evolve into either of:

$$\psi' = \text{| no gas released>}_{\text{device}}\text{| alive>}_{\text{cat}}$$

or

$$\psi' = \text{|gas released>}_{\text{device}}\text{|dead>}_{\text{cat}}, \tag{I.14}$$

for this would mean that one of the terms of the state dropped away. The linearity of the Schrödinger equation does not allow for this.

The Schrödinger equation may be expressed compactly in the following form:

$$\hat{H}\psi = -i\hbar\frac{\partial\psi}{\partial t} \tag{I.15}.$$

This equation describes the evolution of the system's wave function (ψ) over time.[19] On the right-hand side, we find i, $\sqrt{-1}$, and the symbol $\hbar$, which is just Planck's constant, h, divided by 2π.

It may help to view this equation as analogous in some ways to Newton's law: $F = ma$, which may be also written as:

$$F = m\frac{\partial^2 x}{\partial t^2}. \tag{I.16}$$

Newton's law gives us a way to describe the evolution of the position of a material object over time. If we know the object's mass m and the total amount of force F on the object, then we can derive its motion, the change in the value x over time. Likewise in Schrödinger's equation, if we know $\hat{H}$, the system's Hamiltonian, a way of writing down its total energy, then we can derive the change in its wave function ψ over time. To use Newton's equation, one must know something about the system under consideration to be able to write down the forces acting on it. For example, one must ask: are we talking about a ball being dropped

---

[19] More generally, the equation describes the evolution of the state of a system in any basis, not just that of a wave function written out in a position basis.

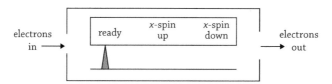

Figure I.13

off a building subject to the force of gravity? Or are we dealing with an asteroid moving in relatively empty space? Once we know this, we can write down which forces are acting on our object. Then if we know what its initial position was, we can calculate its position at any future time using the law. Similarly, to use the Schrödinger equation and predict the evolution of the wave function over time, one must know something about the system under consideration. Are we discussing a particle bouncing back and forth in an otherwise empty box? Or are we talking about an electron circling a nucleus? And so on. Once we know what kind of system we are talking about, we can calculate the system's total energy by writing down a Hamiltonian for the system. The Hamiltonian is an expression that allows us to calculate the total energy of the system across the possible states it might possess. Once we do this, if we know a system's initial wave function, then we can calculate its wave function for any future time using the law.

3 The Measurement Problem

We have now described how quantum mechanics represents states of systems, the connection between quantum states and the statistics of measurement, and finally the dynamical equation of motion. It is now important to see how putting together all of these things produces a paradox—what is commonly referred to as "the measurement problem." I present the problem in the way that has become more or less canonical following the publication of David Albert's *Quantum Mechanics and Experience* (1992). The paradox lies in a conflict between the predictions of the dynamics of quantum mechanics and simple facts about observation.

Say we want to measure an electron's spin in the x-direction. For this purpose, we will use a particular device that we may call "the x-spin detector." Consider an x-spin detector with an opening where electrons may enter and a pointer that can point to one of three labels: "ready," indicating the system is ready to take a new measurement; "x-spin up," indicating the system it has measured has a spin state of up in the x-direction; and "x-spin down," indicating the system it has measured it has a spin state of down in the x-direction (see figure I.13).

If this is a working x-spin measuring device, then when the machine begins in the ready state, after the measurement it will always read the correct state of the incoming particle and this will not affect the spin state of the particle. That is, if the machine is in the ready position and receives an electron with x-spin up, then the machine-particle system will evolve into a state in which the machine's pointer will read "x-spin up" and the electron will remain in the x-spin up state. This may be summarized as the following.

Characteristics of an x-spin detector:

$$|\text{ready}\rangle_m|x\text{-spin up}\rangle_e \rightarrow |\text{ "}x\text{-spin up"}\rangle_m|x\text{-spin up}\rangle_e. \tag{I.17}$$

And:

$$|\text{ready}\rangle_m|x\text{-spin down}\rangle_e \rightarrow |\text{ "}x\text{-spin down"}\rangle_m|x\text{-spin down}\rangle_e. \tag{I.18}$$

This says that if the device (m) is initially in the ready state and if an x-spin down electron (e) enters the detector, then the electron will end up in the x-spin down state and the device will end up pointing to "x-spin down."

What has been described so far are just the conditions on our device being a good x-spin detector.[20] Now suppose that instead of feeding the x-spin detector a particle known to have either x-spin up or x-spin down, we feed it an electron that we have just measured to have up-spin in the z-direction. We can ask what quantum mechanics and the facts we know about x-spin detectors will tell us will happen if we feed this z-spin up electron through our x-spin detector.

As we mentioned in section 2, a fact predicted by quantum mechanics and observed is the following (equation I.5):

$$|z\text{-spin up}\rangle_e = \frac{1}{\sqrt{2}} \; |x\text{-spin up}\rangle_e + \frac{1}{\sqrt{2}} \; |x\text{-spin down}\rangle_e.$$

That is to say, an electron in an up state of z-spin may be rewritten as a super-position of up and down x-spin states. We can then infer that the initial state of a system consisting of the z-up electron and our x-spin detector in the ready state is:

$$|\text{ready}\rangle_m|z\text{-spin up}\rangle_e = |\text{ready}\rangle_m \left\{ \frac{1}{\sqrt{2}} \; |x\text{-spin up}\rangle_e + \frac{1}{\sqrt{2}} \; |x\text{-spin down}\rangle_e \right\}. \tag{I.19}$$

Then, we may distribute to get the following as the initial state of our system before x-spin measurement:

$$|\text{ready}\rangle_m|z\text{-spin up}\rangle_e = \frac{1}{\sqrt{2}} \; |\text{ready}\rangle_m|x\text{-spin up}\rangle_e + \frac{1}{\sqrt{2}} \; |\text{ready}\rangle_m$$
$$|x\text{-spin down}\rangle_e. \tag{I.20}$$

---

[20] See Schrödinger 1935, pp. 158–59.

It then follows from what we know about our detector (from equations I.17 and I.18) and the linearity of the central dynamical law of quantum mechanics, the Schrödinger equation, that our system will evolve in the following way:

$$\frac{1}{\sqrt{2}} \ |\text{ready}\rangle_m |x\text{-spin up}\rangle_e + \frac{1}{\sqrt{2}} \ |\text{ready}\rangle_m |x\text{-spin down}\rangle_e \rightarrow$$

$$\frac{1}{\sqrt{2}} \ | \ ``x\text{-spin up}"\rangle_m |x\text{-spin up}\rangle_e + \frac{1}{\sqrt{2}} \ | \ ``x\text{-spin down}"\rangle_m |x\text{-spin down}\rangle_e. \qquad (I.21)$$

What we are left with, as the final state of our electron/measuring device system, is this:

$$\Psi_{m,e} = \frac{1}{\sqrt{2}} \ | \ ``x\text{-spin up}"\rangle_m |x\text{-spin up}\rangle_e +$$

$$\frac{1}{\sqrt{2}} \ | \ ``x\text{-spin down}"\rangle_m |x\text{-spin down}\rangle_e. \qquad (I.22)$$

This is the state our system evolves into as a result of measurement by our x-spin detector. This raises the question: has anyone ever observed a measuring device/particle system in such a state? What would it look like to observe a measuring device in some superposition of pointing in two places?

What is instead observed is not (I.22) but the following:

either $|``x\text{-spin up}"\rangle_m |x\text{-spin up}\rangle_e$ (with frequency approximately ½)

or $| ``x\text{-spin down}"\rangle_m |x\text{-spin down}\rangle_e$ (with frequency approximately ½). (I.23)

It is (I.22) and not (I.23) that is predicted by the dynamics. The equations say that systems evolve into (I.22). But states like (I.22) are never observed, only states like (I.23). This is the measurement problem. We see the pointer pointing to "x-spin up" or pointing to "x-spin down." We do not see it in a superposition of pointing to "x-spin up" and "x-spin down."

The trouble is that what the laws of quantum mechanics predict is very strange and, more importantly, appears to be in conflict with what we actually observe when we take measurements (e.g., of the x-spin of an electron known to be in an up state of z-spin). Anyone who wants to understand quantum mechanics as a theory of our world (i.e., anyone who wants to be a realist about this theory) therefore must do something to reconcile what the theory predicts with what we observe. To address this conflict, physicists and philosophers have explored several strategies. Some of these are interpretive—leave the theory as it is, but provide a story as to why this doesn't really conflict with observation. Other solutions to the measurement problem involve changing basic details of the theory itself. In the next section, we begin by describing what is often referred to as the orthodox resolution of the problem that involves adding an additional dynamical law to quantum mechanics.[21]

Before ending this section, I would like to briefly return to the Schrödinger's cat thought experiment. The measurement problem is often conflated with the issue raised by Schrödinger, and I would like to make clear that there is a distinction here. There is a way of understanding the cat scenario so that it does present a version of the measurement problem. Here the idea is that we let the cat stay in the opaque box for an hour with the diabolical device. Then we can predict that the state of the whole system will evolve into equation (I.12):

$$\psi = \frac{1}{\sqrt{2}} \; |\text{no gas released}\rangle_{\text{device}} |\text{alive}\rangle_{\text{cat}} + \frac{1}{\sqrt{2}} \; |\text{gas released}\rangle_{\text{device}} |\text{dead}\rangle_{\text{cat}}.$$

So far, there is no measurement problem raised. But then we can ask: what happens if we open the box and look inside? What will we see? The plausible answer is that we won't observe what is described by equation (I.12) but instead will observe:

$$\psi = | \text{no gas released}\rangle_{\text{device}} |\text{alive}\rangle_{\text{cat}} \text{ with probability } \tfrac{1}{2}$$

or

$$\psi = | \text{gas released}\rangle_{\text{device}} |\text{dead}\rangle_{\text{cat}} \text{ with probability } \tfrac{1}{2}. \qquad (I.24)$$

Here we have reached a version of the measurement problem.

However, this is not what I read Schrödinger doing when he presents the cat scenario. Rather, what he presented is more closely related to what is often called the *macro-object problem*. The issue is this: Quantum mechanics, as we've seen, was developed for explaining facts about microscopic systems like electrons in an atom. In developing this theory, it was necessary to conclude that systems could enter into surprising or at the very least interesting states like in figure I.11, states in which the natural thing to say is that the electron is smeared out over a region of space. What Schrödinger did in his thought experiment was to show that by the process of entanglement, such smearing could not be confined to microscopic systems, systems where we might just shrug our shoulders and say, "Okay, I guess electrons can be smeared out like a wave like that," or, if we want to be more like Bohr and Heisenberg, "Well there just isn't any fact about

---

[21] As David Wallace has pointed out to me, although this view is often referred to as the "orthodox" view among philosophers, it is unclear whether it is indeed the working orthodoxy among physicists. It does appear to have an influence on what is presented in introductory-level physics texts, but whether physicists take this seriously is another matter altogether. Just to have a name for what we are about to discuss here, I continue to refer to this as orthodox quantum mechanics.

where the particle is in a case like that." What we are considering in equation (I.12) is a macroscopic system, so both of these responses seem very perplexing. This is one way of understanding the macro-object problem—the problem is to explain how the wave function of a system could ground facts about macroscopic objects. To put it another way, the problem is how a macroscopic system like a cat could either be fundamentally a smeared-out system like that described in (I.12) or the kind of system that it makes no sense to ask questions about its position.

4 Solving the Measurement Problem

There is a (literally) textbook solution to the measurement problem. This results in a modification to quantum mechanics. This modified version of quantum mechanics, orthodox quantum mechanics, was first formulated clearly by Paul Dirac in his 1930 *The Principles of Quantum Mechanics*, and subsequently by John von Neumann in his 1932 text, *The Mathematical Foundations of Quantum Mechanics*.[22] The strategy is to claim that the laws of quantum mechanics, e.g., the Schrödinger equation, are not supposed to be an exhaustive description of how all processes in nature behave. Instead, we must always recognize a distinction between the quantum world and its behavior, and situations where a quantum system is interacting with an external environment, a set of classical devices or creatures that are measuring or observing the quantum system. Von Neumann's proposal was that there are two kinds of processes in nature. He called the inter-action between a quantum system and something external to the system, for example, a measuring device, "Process 1." This is a stochastic, nonlinear process. Process 2 is the deterministic, linear evolution of a quantum system according to the Schrödinger equation.

Process 1 is what is often called wave function collapse, or reduction of the wave packet. We can think of this proposal as adding an additional stochastic law about collapse to the previous dynamics: the collapse postulate. Here, systems collapse onto states having determinate values of a variable with probabilities given by the Born rule. We thus have a resolution of the measurement problem. When evolving on their own, quantum systems may follow Process 2 and there-fore may evolve into states like (I.22):

$$\psi_{m,e} = \frac{1}{\sqrt{2}} \mid \text{``x-spin up''}\rangle_m \vert x\text{-spin up}\rangle_e + \frac{1}{\sqrt{2}} \mid \text{``x-spin down''}\rangle_m \vert x\text{-spin down}\rangle_e.$$

However, once a measurement of x-spin takes place, Process 1 occurs and the wave function (i.e., quantum state) is no longer spread out between the values x-spin up and x-spin down, but collapses onto one or the other, so that it becomes:

---

[22] In what follows, I focus on von Neumann's particular way of formulating the position. For more on orthodox quantum mechanics, I recommend Barrett (1999), chapter 2.

either $|\text{“}x\text{-spin up”}\rangle_m|x\text{-spin up}\rangle_e$ (with probability ½)

or $|\text{“}x\text{-spin down”}\rangle_m|x\text{-spin down}\rangle_e$ (with probability ½),   (I.25)

which, as noted, is what is observed.

In such a way, we can resolve the seeming conflict between the laws of quantum mechanics and experiment. If orthodox quantum mechanics is correct, then all processes in the world may be divided into two fundamental types—Process 1 (in which a measurement is taking place) and Process 2 (in which a measurement is not taking place). If this is correct, there must be some objective fact about which processes in nature count as measurements and which do not.

But is this correct? In the case considered in the previous section, when does this measurement occur? Where does Process 2 end and Process 1 begin? If orthodox quantum mechanics is meant to give us a genuine, objective, exhaustive account of the fundamental physical processes in nature, there must be some answer. Does the collapse of the wave function occur when the electron physically makes its way through the device? When the pointer first begins to move? Only later when some human experimenter observes the device? If the latter, then when exactly in the process of observation does the collapse occur? When light from the device first hits the experimenter's retina, when it gets processed by the visual centers in the brain? When the experimenter first has the conscious thought, "The pointer is reading x-spin up (or x-spin down)"?

John Bell puts the case against orthodox quantum mechanics clearly:

> What exactly qualifies some physical systems to play the role of "measurer"? Was the wavefunction of the world waiting to jump for thousands of millions of years until a single-celled living creature appeared? Or did it have to wait a little longer, for some better qualified system ... with a Ph.D.? If the theory is to apply to anything but highly idealized laboratory operations, are we not obliged to admit that more or less "measurement-like" processes are going on more or less all the time, more or less everywhere? Do we not have jumping then all the time?
>
> The first charge against "measurement," in the fundamental axioms of quantum mechanics, is that it anchors there the shifty split of the world into "system" and "apparatus." A second charge is that the word comes loaded with meaning from everyday life, meaning which is entirely inappropriate in the quantum context. (1987, p. 216)

Bell's worries appear to be (1) that there is no clear distinction between when there is a measurement and when there is not, and (2) that the word *measurement* has inappropriate connotations (see also Schrödinger 1935, p. 158). Focusing on (2) first, Bell says, "When it is said that something is 'measured' it is difficult not to think of the result as referring to some pre-existing property of the object in question" (1987, p. 216). Whereas in quantum mechanics it is supposed to be the case that Process 1 (the wave function collapse) at least often involves quantities that are indeterminate coming to have determinate values, the use of the word *measurement* to describe this process makes it seem as if the system had these determinate values even before the collapse.

Eugene Wigner is one physicist who took the first set of concerns about the slipperiness of the notion of measurement seriously early on. In his 1961 article, "Further Remarks on the Mind-Body Question," Wigner suggested that we interpret "measurement" in a precise way to mean interaction of a physical system with an irreducible consciousness. His stipulation does appear to result in a precise demarcation of systems into those following Process 1 and Process 2 so long as there is a precise and objective distinction about where there is consciousness and where there is not. However, this assumes a fundamental duality between consciousness and physical systems that many contemporary philosophers will find unsatisfactory given the widespread assumption of physicalism.

Bell (1987, p. 216) anyway raises a more fundamental worry about orthodox quantum mechanics. This is that by introducing the concept of measurement into the basic laws of quantum mechanics, we make this theory about measurers or observers, about us, in a way that is inappropriate if this is to be a true, fundamental physical theory of the world. A fundamental theory should not make appeal to such complex organisms as observers as part of its most basic framework.

One physicist who clearly recognized the problem with the orthodox formulation of quantum mechanics was David Bohm. After writing a definitive textbook[23] explaining the orthodox formulation, the next year (1952), Bohm published an alternative to it. Bohm's theory is the successor to a theory first proposed by de Broglie in 1927 at the Solvay Congress, though Bohm notes that when he initially completed his theory, he wasn't yet aware of de Broglie's proposal. Bohm took the fundamental trouble with quantum mechanics to be that it assumes both (1) that all the quantum state (the wave function) can give one is a statistical claim about what is likely to happen as a result of measurement, and (2) that the wave function gives a complete account of a system. He argues:

> The assumption that an individual system is completely described by the wavefunction and its probability interpretation implies a corresponding unavoidable lack of precision in the view's conceptual structure, with the aid of which we can think about and describe the behavior of the system. (1952, p. 371)

His worry is as follows. Let's say we take a measurement of an electron to find its position in a case where before the measurement is performed the quantum state

---

[23] This is Bohm (1951).

of the system is correctly described as being in a superposition of position states. According to orthodox quantum mechanics and the standard interpretation, at this time there is no fact about the position of the particle. Indeed, according to Bohr's principle of complementarity, the concept of position does not even apply to the electron at that time. All that can be said is that if we were to take a measurement, the quantum state would collapse onto one result or another with certain probabilities. Bohm's complaint is that quantum mechanics then fails to give us an account of why and how phenomena at the microscopic level occur. He continues:

> The principle of complementarity states that no single model is possible which could provide a precise and rational analysis of the connections between these phenomena [before and after measurement]. In such a case, we are not supposed, for example, to attempt to describe in detail how future phenomena arise out of past phenomena. Instead, we should simply accept without further analysis the fact that future phenomena do in fact somehow manage to be produced, in a way that is, however, necessarily beyond the possibility of a detailed description. The only aim of a mathematical theory is then to predict the statistical relations, if any, connecting the phenomena. (Bohm 1952, p. 371)

The point Bohm wants to make is that this pessimistic view about what our fundamental physical theories can accomplish (they can only give us probabilities for measurements) may have been justified if orthodox quantum mechanics (or the version presented in Section 2) was the only theory on the table, the only option we had for a fundamental theory. However, he says, there is an alternative, one that has more ambitious goals than merely giving statistics about what could happen.

The strategy Bohm deploys in creating this alternative theory relies on rejecting the assumption that the quantum state is a complete description of reality (that the wave function is everything). In this, he is explicitly influenced by Einstein in the EPR paper. Even in a case where the quantum state of an electron is like the following one we considered in Section 2 (equation I.2):

$$\psi = \frac{1}{\sqrt{2}} \, |3, 0, 0> + \frac{1}{\sqrt{2}} \, |4, 0, 0>.$$

Bohm says there is a definite fact about the position of the electron. (To be consistent with this quantum state, it must be at either (3, 0, 0) or (4, 0, 0).) According to Bohm, particles always have definite positions. So the probabilities contained in the traditional understanding of the quantum state are not fundamental, objective chances. As he puts it, they are merely "pragmatic." They express the limitations of our knowledge perhaps about where the electron is, but they do not express all there is to know about where it is. Bohm's theory is an example of what is referred to as a hidden variables theory—a theory in which more variables are posited than are contained within the quantum state.[24]

In rejecting the assumption that the quantum state expresses all there is fundamentally to know about a system, and positing determinate particle positions at all times, Bohm thereby allowed for a way to solve the measurement problem that did not (as the orthodox account did) involve making the concept of a measurement or observation part of the fundamental laws. In our presentation of the measurement problem we noted that due to the linearity of the Schrödinger equation, systems in superpositions of (say) position never evolve out of these superpositions—what corresponds to terms in the sum do not go away. This caused a problem because we never observe systems in a smeared-out state of being here and there or observe a pointer in a superposition of pointing here and there. Bohm has the resources to explain why this is: because pointers and other objects always have determinate locations. They are never smeared out over locations in the way a naive reading of the quantum state might suggest.

This raises the question, then, of what the quantum state (and hence the Schrödinger dynamics) describes if not the position state of the particles. Here we can return to de Broglie and his original proposal. According to de Broglie, there are fundamentally two kinds of things: particles and waves. The quantum state describes the state of the wave. The role of this wave is to guide the particle into various states. For this reason, this is called the pilot wave model. In 1927, de Broglie rejected this model under criticism from Wolfgang Pauli at the Solvay Congress and a later critique by von Neumann. But Bohm responded to both Pauli's and von Neumann's criticisms in his 1952 paper.[25]

This version of (or alternative to) quantum mechanics is now commonly known as Bohmian mechanics. Bohmian mechanics has received a steady increase in support among philosophers of physics, especially since its elaboration in 1992 in a paper by Detlef Dürr, Sheldon Goldstein, and Nino Zanghì.[26] According to Bohmian mechanics, there are two fundamental dynamical laws. The first, for the nonrelativistic case, is the Schrödinger equation, which describes the deterministic, linear evolution of the wave function over time:

$$\hat{H}\,\Psi = -i\hbar\frac{\partial \Psi}{\partial t}.$$

[24] It is often noted that this is something of a misnomer, making the theory seem more radical than it is because positions are not the kind of variable that are really hidden from us. In the title of his paper, Bohm himself puts "hidden" in quotes. This is a good place to mention that there is another class of hidden variables theories that we will not be discussing in detail here. These have many details in common with Bohm's theory: they are the modal theories. The main difference is in the kind of feature that is said to possess a determinate value. See Healey (1989) and van Fraassen (1991).

[25] To discuss this exchange would take us too far from the main aims of this introduction. For more details, see Cushing (1994), chapters 7 and 8.

[26] A distinct way of elaborating Bohm's 1952 approach is found in the collaborative work of Bohm and Basil Hiley, for example, Bohm and Hiley (1993).

Here I use a capital Ψ to indicate that the law is about the evolution of the universal wave function, that is, the wave function that describes the universe as a whole, and exists as a field on configuration space. The second law is most often referred to as the guidance equation. This law describes how the state of the particles' positions evolve as a function of the state of the wave function at a time. This law is also deterministic. Where Q_k refers to the position of an arbitrary particle:

$$\frac{dQ_k}{dt} = \left(\frac{\hbar}{m_k}\right) \text{Im} \frac{\Psi^* \partial_k \Psi}{\Psi^* \Psi}(Q_1,...,Q_N).^{27}$$

Although both of these laws are deterministic, to capture the statistical predictions of quantum mechanics, one has to make sense of where the Born probabilities come from, and this gets explained ultimately on the basis of our inability to know everything about the positions of particles. See Dürr, Goldstein, and Zanghì (1992) for more details.

One might ask why Bohm and his followers thought it was a sensible idea to privilege particle position in this way above other features of particles. One helpful clue comes from Bell, who once made the observation that "in physics the only observations we must consider are position observations, if only the positions of instrument pointers" (Bell 1987, p. 166). The idea is that if we are trying to make a connection between our theories and observations—which is what we are trying to do, after all, when we are trying to solve the measurement problem—then if we can get the observations of positions right, then the observations of everything else get captured as well. How do we observe anything except by observing the position of something? As we will see, Bohmian mechanics is not the only version of quantum mechanics that privileges position, and it seems to do so for good reason.

It is worth noting that although Bohm does cite Einstein for inspiration, the motivation for adopting this kind of hidden variables approach to quantum mechanics today is not to avoid nonlocality. We know from Bell's work that even a hidden variables approach will be nonlocal. It is often claimed that what Bell showed was precisely that hidden variables approaches are mistaken. Note that this would only be so if the primary motivation for them were to sidestep nonlocality. This is not the motivation—rather, as we already saw, Bohm's motivation is to provide a theory that gives precise laws specifying the evolution of states in all cases and to provide an adequate solution to the measurement problem.

We saw that one of Bohm's stated objectives for introducing this new version of quantum mechanics was to move away from the assumption that the wave function gave a complete description of physical systems. By adding particles with definite positions to his picture, he could give a precise

---

[27] See the chapter by Goldstein and Zanghì for further details on this equation.

account of all processes that did not have the holes in explanation of orthodox quantum mechanics. Moreover, he could do so without building the
imprecise and ambiguous concept of measurement into the fundamental
dynamical laws.

Around this time, other physicists started to see ways to understand quantum mechanics that didn't posit hidden variables and so made the wave function a complete description of reality without the problems Bohm worried about.
We will discuss two such approaches. The first was elaborated five years after
Bohm's hidden variables paper by Hugh Everett.[28] Everett's idea was to reject the
claim that the Schrödinger dynamics needed to be modified or supplemented.
The challenge, then, was to explain why if this is so, quantum mechanics does
not conflict with our observations. Everett's paper "'Relative State' Formulation
of Quantum Mechanics" states:

> This paper proposes to regard pure wave mechanics (Process 2) as a
> complete theory. It postulates that a wave function that obeys a linear
> wave equation everywhere and at all times supplies a complete math-
> ematical model for every isolated physical system without exception.
> (1957, p. 316)

What, then, to say about our observation of pointers in definite positions?

To see Everett's solution, let us go back to the earlier example of a z-spin up
electron passing through an x-spin detector. We noted that this system will be
expected, according to the wave dynamics, to pass into the state (I.22):

$$\Psi_{m,e} = \frac{1}{\sqrt{2}} \ |\text{``}x\text{-spin up''}>_m|x\text{-spin up}>_e + \frac{1}{\sqrt{2}} \ |\text{``}x\text{-spin down''}>_m|x\text{-spin down}>_e,$$

and this conflicts with what we would actually observe if we ran this experiment.
Orthodox quantum mechanics would say that at the time of measurement, this
system actually collapses out of the state described in (I.22) into (I.25):

either $|\text{``}x\text{-spin up''}>_m|x\text{-spin up}>_e$ (with probability approximately ½)

or $|\text{``}x\text{-spin down''}>_m|x\text{-spin down}>_e$ (with probability approximately ½),

which is what we would observe.

Everett says that this is incorrect. The wave function never collapses in this way
(there is no Process 1). Instead, we must recognize that the *total* wave function of
the electron/detector system is indeed as (I.22) describes. "So far as the complete
theory is concerned all elements of the superposition exist simultaneously, and
the entire process is quite continuous" (1957, p. 318). We can look at (I.22) as a

---

[28] I do not mean to imply here that Everett knew about or was responding to Bohm's work.

superposition of pairs of electron/detector states in which each of these corresponds to something we might observe as a result of measurement:

> There is a representation in terms of a *superposition*, each element of which contains a definite observer state and a corresponding system state. Thus with each succeeding observation (or interaction), the observer state "branches" into a number of different states. Each branch represents a different outcome of the measurement.... All branches exist simultaneously in the superposition after any given sequence of observations. (Everett 1957, p. 320)

Everett's distinction between the total state of the system (what is the complete state of the system; its wave function evolving deterministically and linearly according to the wave mechanics) and the parts of that state that are the results of measurements is the key to his proposed reconciliation of the quantum formalism and actual observations. When an observation takes place, the system will evolve into:

$$\Psi_{o,m,e} = \frac{1}{\sqrt{2}} \; |\text{belief: } x\text{-spin up}\rangle_o |\text{"}x\text{-spin up"}\rangle_m |x\text{-spin up}\rangle_e + $$

$$\frac{1}{\sqrt{2}} \; |\text{belief: } x\text{-spin down}\rangle_o |\text{"}x\text{-spin down"}\rangle_m |x\text{-spin down}\rangle_e. \qquad (I.26)$$

Then, there will be observers (o) seeing the detector registering x-spin up. There will also be observers seeing the detector registering x-spin down. These branches both exist, Everett says. The question is how this is supposed to make sense. Why don't we notice both pointer readings?

A common way of viewing Everettian quantum mechanics is to say that in an act of measurement, the universe splits into two. There is a world in which the electron has x-spin up, the pointer points to "x-spin up," and we believe the electron has x-spin up. There is another world in which the electron has x-spin down, the pointer points to "x-spin down," and we believe the electron has x-spin down. This is why Everettian quantum mechanics is often called "the many worlds interpretation." Because the contrary pointer readings exist in different universes, no one notices that both are read. This way of interpreting Everettian quantum mechanics raises many metaphysical difficulties. Does the pointer itself split in two? Or are there two numerically distinct pointers? If the whole universe splits into two, doesn't this wildly violate conservation laws? There is now twice as much energy and momentum in the universe than there was just before the measurement. How plausible is it to say that the entire universe splits?

Although this "splitting universes" reading of Everett is popular (Deutsch 1985 speaks this way in describing Everett's view, a reading originally due to Bryce Dewitt), fortunately, a less puzzling interpretation has been developed. The idea is to read Everett's theory as he originally intended. Fundamentally, there is no splitting, only the evolution of the wave function according to the Schrödinger dynamics. To make this consistent with experience, it must be the case that there are in the quantum state branches corresponding to what we observe. However, as, for example, David Wallace has argued (2003, 2010), we need not view these branches—indeed, the branching process itself—as fundamental. Rather, these many branches or many worlds are patterns in the one universal quantum state that emerge as the result of its evolution. Wallace, building on work by Simon Saunders (1993), argues that there is a kind of dynamical process; the technical name for this process is "decoherence," that can ground the emergence of quasi-classical branches within the quantum state. Decoherence is a process that involves an interaction between two systems (one of which may be regarded as a system and the other its environment) in which distinct components of the quantum state come to evolve independently of one another. That this occurs is the result of the wave function's Hamiltonian, the kind of system it is. A wave function that (due to the kind of state it started out in and the Schrödinger dynamics) exhibits decoherence will enter into states capable of representation as a sum of noninteracting terms in a particular basis (e.g., a position basis). When this happens, the system's dynamics will appear classical from the perspective of the individual branches.

So, for example, Everettians of this stripe will argue the quantum state of our universe is such that it can be represented as a superposition of terms, each of which ground classical-looking macro-objects. And since also as a result of decoherence, interference between the terms washes away, we will not observe other parts of the quantum state even though they are there. We thus get a resolution of the Schrödinger macro-object problem. The quantum state doesn't describe one smeared-out cat, but instead two cats, one of which is dead and one of which is alive.

Note that facts about the quantum state decohering are not built into the fundamental laws. Rather, this is an accidental fact depending on the kind of state our universe started out in. The existence of these quasi-classical states is not a fundamental fact either, but something that emerges from the complex behavior of the fundamental state. The sense in which there are many worlds in this way of understanding Everettian quantum mechanics is therefore not the same as it is on the more naive approach already described. Fundamentally there is just one universe evolving according to the Schrödinger equation (or whatever is its relativistically appropriate analog). However, because of the special way this one world evolves, and in particular because parts of this world do not interfere with each other and can each on their own ground the existence of quasi-classical macro-objects that look like individual universes, it is correct in this sense to say (nonfundamentally) there are many worlds.

There is another alternative to orthodox quantum mechanics that, like Everettian many worlds theories and unlike Bohmian mechanics, views the quantum state as a complete description of reality. This approach is more similar to orthodox quantum mechanics in that, unlike Everettian quantum mechanics and Bohmian mechanics, it posits a collapse of the wave function. In doing so, this approach can explain our observation of, for example, determinate pointer readings without positing hidden variables. It also does not resort to positing the existence of other worlds we cannot observe (be they nonfundamental). We saw that the problem with the orthodox approach was that to explain when collapse of the wave function occurred (when the quantum state moved from being indeterminate between position, momentum, or spin states to determinate), it appealed to the notion of measurement. This seemed to be inappropriate in the context of a fundamental physical theory. The notion of measurement is imprecise, and measurements (whatever they are) seem too complex to be apt subjects of fundamental physical law.

It would be an improvement, however, if there were a natural mechanism by which the quantum state would collapse spontaneously, especially if it collapsed in the cases we think it does and onto the kind of states we think it does. For example, in the case we used to illustrate the measurement problem, it would be an improvement if the laws of quantum mechanics were such that the state of that system collapsed sometime before the time we looked at the x-spin detector and collapsed onto a state in which the pointer was determinately pointing at "x-spin up" or "x-spin down."

This is exactly how the 1986 approach of GianCarlo Ghirardi, Alberto Rimini, and Tullio Weber (GRW) works. As Bell summarizes the approach:

> The idea is that while a wave function … normally evolves according to the Schrödinger equation, from time to time it makes a jump.… The probability per unit time for a GRW jump is N/τ where N is the number of arguments in the wave function and τ is a new constant. The jump is to a "reduced" or "collapsed" wavefunction. (Bell 1987, pp. 202–3)

GRW actually posit two new constants in nature. One, as Bell notes, is a constant τ, that relates to how often there is a jump in the quantum state. To make the theory match with experiment, this constant will be large enough so that it is very unlikely that a single particle system will undergo a collapse in a given unit of time. As the complexity of the wave function increases, the probability of a jump increases. For example, we can discuss what will happen when we have a system consisting of a single electron interacting with a z-spin measuring device with a wave function like equation (I.22):

$$\psi_{m,e} = \frac{1}{\sqrt{2}} \ |\text{"x-spin up"}\rangle_m |x\text{-spin up}\rangle_e + \frac{1}{\sqrt{2}} \ |\text{"x-spin down"}\rangle_m |x\text{-spin down}\rangle_e.$$

The way $\psi_{m,e}$ is written here is deceptive, of course. Each term is in fact much more complex than presented because it takes a very complex system of microscopic systems to make an x-spin measuring device that is pointing to "x-spin up" or "x-spin down." For example, pointers themselves will be made of vast numbers of particles. Due to the complexity of this wave function, it will be much more likely for a jump to occur according to GRW.

This ensures that collapses (jumps) occur when we expect them to. There is another issue about how the theory ensures that collapses occur how we expect them to. What we are looking for in our case is that quantum states jump onto states in which, for example, our pointers are determinately pointing one way or another. Like Bohmian mechanics, GRW does this by focusing on getting the positions right. Quantum jumps are jumps onto a collapsed wave function with the following form:

$$\psi(t, r_1, r_2, \ldots, r_n) \rightarrow \psi'(t, r_1, r_2, \ldots, r_n) = \frac{j(x - r_n)\psi(t,\ldots)}{R_n(x)} . \qquad (\text{I.27})$$

The dynamics takes in a wave function expressible as a sum in a position basis. The initial wave function is in effect multiplied by a so-called jump factor, $j(x)$, which localizes the wave function around one of the terms in the initial state. GRW represent the jump factor as:

$$j(x) = Ke^{-x^2/2a^2}, \qquad (\text{I.28})$$

where a is the second new constant. Which position term survives the collapse is determined randomly on this theory, with (I.27) predicting statistics closely matching the predictions of the Born rule. For this reason, GRW is supposed to mirror the dynamics of the orthodox approach in many ways. It improves on that approach by making collapse of the wave function a natural, spontaneous process, rather than something arising due to the interaction of a system with an observer or measuring device.

5 Matters of Ontology

We have now outlined several different approaches to quantum mechanics. All but Bohmian mechanics work with the assumption that the wave function is all one needs to specify to know the complete and fundamental state of a system. The Everettian and several collapse approaches we considered assume that all there is fundamentally is the wave function, and the law(s) of quantum mechanics specify its evolution over time (either linear and deterministic or nonlinear and stochastic).

As metaphysicians, we are interested in the question of what the world is fundamentally like according to quantum mechanics. Some have argued that the answer these accounts give us (setting aside Bohmian mechanics for the moment) is that fundamentally all one needs to believe in is the wave function. What is the wave function? It is something that as we have already stated may be described as a field on configuration space, a space where each point can be taken to correspond to a configuration of particles, a space that has $3N$ dimensions where N is the number of particles. So, fundamentally according to these versions of quantum mechanics (orthodox quantum mechanics, Everettian quantum mechanics, spontaneous collapse theories), all there is fundamentally is a wave function, a field in a high-dimensional configuration space. The view that the wave function is a fundamental object and a real, physical field on configuration space is today referred to as "wave function realism." The view that such a wave function is everything there is fundamentally is wave function monism.

To understand wave function monism, it will be helpful to see how it represents the space on which the wave function is spread. We call this space "configuration space," as is the norm. However, note that on the view just described, this is not an apt name because what is supposed to be fundamental on this view is the wave function, not particles. So, although the points in this space might correspond in a sense to particle configurations, what this space is fundamentally is not a space of particle configurations. Likewise, although we've represented the number of dimensions configuration space has as depending on the number N of particles in a system, this space's dimensionality should not really be construed as dependent on the number of particles in a system. Nevertheless, the wave function monist need not be an eliminativist about particles. As we have seen, for example, in the Everettian approach, wave function monists can allow that there are particles, derivative entities that emerge out of the decoherent behavior of the wave function over time. Wave function monists favoring other solutions to the measurement problem can also allow that there are particles in this derivative sense. But the reason the configuration space on which the wave function is spread has the number of dimensions it does is not, in the final analysis, that there are particles. This is rather a brute fact about the wave function, and this in turn is what grounds the number of particles there are.[29]

There are a couple of issues here. The first is answering the question: what is it about this field in $3N$-dimensional configuration space that allows it to ground the existence of a multiparticle system in three-dimensional space? There are deep conceptual issues here and some, for example, John Hawthorne (2010), have suggested that the wave function monist faces an explanatory gap similar to what the physicalist philosopher of mind faces in trying to explain how physical facts can ground consciousness. Just as many physicalists are optimistic about closing the explanatory gap in the case of consciousness, here too, many wave function monists are optimistic about closing this explanatory gap. In both cases, the strategy is to appeal to functionalism. There are particles in a three-dimensional space just in case the wave function behaves in such a way over time as to play the causal role of a system of N particles in a three-dimensional space.[30]

---

[29] Which, as Wallace (2010) and Albert and Loewer (1996) both emphasize, might turn out to be vague.

[30] See Albert's contribution to this volume.

Once we have justification for saying that there are N particles in a (derivative) three-dimensional space, we can ask which features of this wave function ground facts about the different particles being in particular locations. In a paper on GRW (and collapse theories more generally), Albert and Barry Loewer suggest the following rule: "Particle x is in region R" is true iff the proportion of the total squared amplitude of x's wave function which is associated with points in R is greater than or equal to $1 - p$, where p is some real number between 0 and 1/2 (1996, p. 87). They argue that there isn't any precise fact about what p is and so there isn't always a precise answer about the locations of particles. This imprecision is a feature of the Everettian's decoherence-based account as well, but doesn't seem to be a problem. The fundamental wave function ontology is precise in both cases.

Wave function monism (and realism more generally) is the topic of a great deal of discussion throughout the present volume. It is not, however, a plausible way to understand the ontological ramifications of Bohmian mechanics because this view is motivated primarily as a way to get away from the assumption that the quantum state can give a complete representation of systems. The natural way to understand the ontology of Bohmian mechanics is to see it as saying that there are fundamentally two kinds of entities. There is the wave function, whose evolution over time is described by the Schrödinger equation. Then there are particles, whose behavior depends (as is clear in the guidance equation) on the state of the universal wave function at a given time. These particles always possess determinate locations in three-dimensional space. The question is then raised: what is the ontological status of the wave function on an approach like this? Here, there appear to be several options. One could in principle attempt to understand the wave function in the same way as we proposed for Everettian quantum mechanics and the collapse approaches: say the wave function is a field in configuration space (which isn't really a space of configurations but again is itself a genuine, physical space).[31] However, this only introduces questions about how the behavior of the particles in ordinary three-dimensional space may be guided by a field in an altogether different space. A more common approach to Bohmian mechanics deploys a different way of understanding the wave function, viewing it as something more akin to a law that is able to determine the motion of the particles.[32] On such an interpretation, there is only one genuine physical space—the space in which the particles move. Even though the wave function is a genuine entity, those who take this view reject categorization as wave function realists, because even though the wave function is real, it is not taken to be a genuine, physical field. The wave function cannot be eliminated from the ontology because, as we illustrated using the example of entanglement

---

[31] See Gordon Belot (2012) for a discussion of this option (and several others) for the Bohmian.

[32] See the contributions from Valia Allori, Sheldon Goldstein, and Nino Zanghì in this volume, as well as Allori, Goldstein, Tumulka, and Zanghì (2007).

of position, there is more to the quantum state than is carried in the states of the particles themselves. However, as these authors say, the wave function is not an element of the theory's primitive ontology, what the theory is fundamentally about. What the theory is fundamentally about is a configuration of particles in ordinary space.[33]

So far we have considered what are in a sense the most straightforward interpretations of the various solutions to the measurement problem: wave function monist interpretations of Everettian quantum mechanics and collapse theories, and two interpretations of Bohmian mechanics. This may make it seem as if the central ontological issues in quantum mechanics ultimately boil down to the issue of which way of solving the measurement problem is correct, and in particular whether a solution to the measurement problem ought to be a hidden variables theory. Things are a bit more subtle, however, and the easiest way to see this is to understand that there are ways to be a Bohmian and endorse wave function realism and there are ways to prefer GRW or Everettian quantum mechanics while denying wave function monism.

Let's start with the latter. There are many different reasons why one might opt for a version of quantum mechanics with a fundamental ontology that does *not* simply contain a wave function in configuration space. In section 1 we discussed Bohr's correspondence principle: the insistence that whatever weirdness there was to be found in the nature of quantum states, there had to be a way to line up the predictions of the theory with classical predictions for ordinary macroscopic objects. Bohr must be basically right about this—there has to be a way of connecting quantum mechanical predictions with what we know to be the behavior of ordinary macroscopic objects. Following some remarks by Bell (1987, p. 52), some have thought that this connection with ordinary, classical objects is perhaps best placed right into the fundamental dynamical equations rather than confining it to what Bell called the "surrounding talk" of the theory, such as rules about how to connect quantum predictions with experience.

Bell had a particular suggestion about how to do this: one should make sure physical theories had what he called "local beables." By "beables," he just meant an entity, something that *is*.[34] Local beables are those that don't just exist, but exist somewhere in ordinary three-dimensional space or four-dimensional space-time (Bell 1987, p. 53). Philosophers and physicists have offered different reasons for why physical theories should have local beables. Bell himself said that theories were required to have local beables if they were to allow for any cases of local causality (1987). One can't address the issue of whether interactions in a theory are taking place locally or superluminally across spatiotemporal distances if the

---

[33] For more on the notion of a theory's primitive ontology, see Allori's contribution to this volume. The notion of a theory's primitive ontology is distinct from the notion of a theory's fundamental ontology.

[34] He explicitly contrasted *beables* with Heisenberg's *observables*.

theory makes no reference to entities that have spatiotemporal location. Maudlin (2007) makes the point that a theory needs local beables if there is to be a clear way in which we may see it as being confirmed. Roughly, the idea is that for there to be confirmation, there must be a class of entities for which both (1) the theory makes claims about them, and (2) we have some way of observing them (even if indirectly) and hence confirming these claims. But our observations are always of objects with locations in ordinary space. We see the pointer here, see the scintillation pattern on the screen there, and so on. Because our observations are of local beables, justified theories must make claims about local beables. The wave function isn't a local beable because it has no location in ordinary space or space-time. So we shouldn't accept wave function monist versions of quantum mechanics.

One option for those who believe that quantum mechanics should have local beables is to endorse Bohmian mechanics over other solutions to the measurement problem. However, Bohmian mechanics is not the only version of quantum mechanics that posits local beables in addition to the wave function. Recently, alternative versions of both GRW and the many worlds theory have been developed. Just as Bohmian mechanics has two laws, one describing the evolution of the wave function over time, the other describing the way the wave function guides the evolution of the particles, these alternative versions of GRW and many worlds theory contain additional laws relating the behavior of the wave function to the evolution of local beables: objects in ordinary space-time. Two alternative versions of GRW have been proposed: GRW_m, which adds a law relating the wave function to a field of mass spread out with varying density in three-dimensional space (Bassi and Ghirardi 2003),[35] and GRW_f, which adds a law relating the wave function to additional, fundamental events called "flashes," which take place at the locations corresponding to where the wave function jumps or collapses (Bell 1987, p. 205). As Bell says:

> The GRW jumps … are well localized in ordinary space. Indeed each is centred on a particular spacetime point (x, t). So we can propose these events as the basis of the "local beables" of the theory…. A piece of matter then is a galaxy of such events. (1987, p. 205)

Allori, Goldstein, Tumulka, and Zanghì (2011) have also proposed a version of many worlds theory.[36] The theory they call "S_m," starts with the Schrödinger equation and adds a new law relating the state of the wave function to the evolution of a mass density field on three-dimensional space. The ontology of

---

[35] Although much recent philosophical discussion of GRW_m emphasizes the fact that GRW contains local beables, in the paper cited above, the mass densities get introduced in order to connect GRW to a theory of gravitation.

[36] It is not clear that it is appropriate to call this a version of Everettian quantum mechanics since the spirit of the Everett approach is to keep the wave mechanics (Process 2) as is. The theory S_m proposed in Allori et. al. (2011) adds an additional process to Process 2.

this theory is, like GRW$_m$, a wave function and a mass density field. The difference between the two theories is the way the wave function (and hence mass densities) evolve. Without the collapse mechanism, the regions of high mass density will continue to spread through three-dimensional space on S_m. Allori et al. suggest that this will happen in such a way as to ground the existence of many worlds:

> It is easy to see that S_m has a certain many-worlds character, since if ψ is the wave function of Schrödinger's cat then there will be two contributions to the m function, one resembling a dead cat and the other a live cat.... For now note the duality: there exist two things, the wave function ψ and the matter density function m. The latter represents the "primitive ontology" ..., the elements of the theoretical picture that correspond to matter in 3-dimensional space; the wave function tells the matter how to move. (Allori et al. 2011, p. 4)

On all of these versions of GRW and many worlds theory, where something beyond the wave function is posited, there is a live question regarding the status of the wave function. As we noted, Allori et al. (2007) have suggested that the wave function has a nomological status. However, there are other options. Maudlin, though sympathetic to the idea that one needs to have a version of quantum mechanics that refers to local beables and rejects the picture of the wave function as a field on a genuine, physical configuration space, is less disposed to view it so much as akin to a law of nature. Instead, he suggests that one should recognize the wave function as instancing a novel ontological kind, rather than trying to force it into another ontological category. Taking a different approach, Bradley Monton (2006, this volume) has argued that the wave function is better viewed as a property of quantum systems:

> On my picture of what quantum mechanics is about, quantum mechanics is about particles and systems of particles, all evolving in three dimensional space ... the wave function doesn't exist on its own, but it corresponds to a property possessed by the system of all the particles in the universe (or whatever closed system you're interested in). (2006, p. 779)

The sort of property the wave function corresponds to will then depend on the version of quantum mechanics one prefers. Finally, Wallace and Christopher Timpson (2010), while defending wave function monism from some of the worries noted here and in the rest of this volume, have proposed an alternative view where the quantum state is not properly seen as representing a field in 3N-dimensional space. They argue instead that to make quantum theory consistent with relativity, one should prefer a view they call "spacetime state realism."[37] Here the fundamental objects are space-time regions, which get assigned highly abstract intrinsic properties derived from the Hilbert space formulation. Note that although this proposal, like those just considered, technically involves an ontology of local beables, this is not its motivation. The account is motivated as an alternative to wave function realism based on considerations from relativity.[38]

So there are several alternatives to wave function monism, even if one prefers the Everettian or GRW versions of quantum mechanics. These alternatives are motivated by a variety of factors from a certain view about what is required to ground the existence of macro-objects to relativity theory. Just as some have proposed versions of collapse and many worlds theories that posit additional objects in space or space-time, those that are drawn more to wave function realism have themselves offered a way of understanding Bohmian mechanics so that it (fundamentally) contains no objects in three-dimensional space. Recall that Bohmian mechanics contains two fundamental dynamical equations: the Schrödinger equation, which describes the evolution of the wave function and the guidance equation, which describes the evolution of the particle configuration in terms of the wave function. As Albert has pointed out,[39] it is possible to read both of these equations as being about entities in $3N$-dimensional configuration space. We may start by interpreting the wave function realistically, as a field on configuration space. Then we may interpret what the guidance equation describes not as a configuration of many particles in a separate three-dimensional space, but as one particle representing that whole configuration. According to this view, the guidance equation is about the evolution over time of one world particle.

This interpretation of Bohmian mechanics is more like the pilot-wave model de Broglie originally intended. For here it is much more natural to see the wave function as a wave pushing the particle (the so-called world particle) around because (1) both the particle and the wave function exist in the same manifold of configuration space, and (2) the wave is really a wave, rather than a law or a property.

If one was attracted to Bohmian mechanics because it is a theory with local beables built right in fundamentally, then of course one will not be tempted to read the theory in this way. For example, one might imagine Maudlin pointing out that even if there might be a way to read the locations of the many particles off of the location of the world particle in configuration space—the theory is not ontologically complete (Maudlin 2007, p. 3154). It's not ontologically

[37] One can find more discussion of the putative problems for wave function realism raised by relativity in the last section of this introduction.

[38] Those who emphasize a need for local beables for the reasons presented above usually prefer accounts with local beables that connect in a natural way with the manifest image. So the proposed ontologies are particles with positions or masses. As Wallace and Timpson stress, the fundamental entities they propose are not going to be so familiar.

[39] See his contribution to this volume and Albert (1996).

complete because there aren't any local beables on this reading, or anything else from which we may ground the existence of the ordinary macroscopic objects we use to get confirmation for the theory. So, those who opt for this way of understanding Bohmian mechanics tend to be those who think that the wave function realist does have a way to ground the existence of ordinary macro-objects in the wave function (and a world particle) in configuration space.

This version of Bohmian mechanics has been the target of an important critique by Brown and Wallace (2005). Brown and Wallace point out that if we view the ontology of Bohmian mechanics in the way just suggested, then the only difference between this and Everettian quantum mechanics is that in Bohmian mechanics there is this additional object, the world particle, that is supposed to ground in a certain sense which state is actualized (e.g., living cat or dead cat). Brown and Wallace raise the question of the point of this additional posit. They argue that because the wave function is sufficient to ground the rest of the world (as Everettians take themselves to have effectively argued), the world particle is superfluous and so Bohmian mechanics is just an unparsimonious version of the many worlds theory. This exchange[40] raises important questions about the role of the wave function vis-à-vis the particles in grounding macroscopic objects in Bohmian mechanics. One thing that appears to be left out of the 2005 discussion is that the particle does play the role of grounding what the chances in Born's rule are about on Bohmian mechanics. It is often argued that Everettian models by themselves are unable to ground such chances because all possible outcomes are realized. For example, after the hour passes, there is a living cat and a dead cat. It is unclear what the probabilities in the Born rule are supposed to be about. If the many worlds theory is true, there isn't a 50% chance the radioactive substance will decay. The chance is 100%, and the cat will die with certainty.[41]

The essays collected in this volume take very different views regarding the ontological status of the wave function (whether it is a field in a $3N$-dimensional space, a law, a property of an ensemble of particles or a space-time region, or as Steven French considers in his contribution: a blob). However, all of these alternatives presuppose that ψ represents something, something that is ontologically more fundamental than and independent of us as subjects capable of thinking of it. I conclude this section by noting that not everyone shares this view. There are many physicists and philosophers who have argued that this is to taken an incorrect approach to understanding the wave function. Jeffrey Bub and Itamar Pitowsky (2010) argue that the measurement problem described in section 3 is a pseudo-problem and only arises from the dogma that the quantum state has ontological significance. Physicist Robert Spekkens (2007) has distinguished between what he calls "ψ-ontic" and "ψ-epistemic" views, where all of

---

[40] See further discussion by Valentini 2010 and Brown 2010.
[41] See Greaves 2007.

the views represented in this volume (and all of those represented above) would be ψ-ontic views.[42] Those who take a ψ-epistemic approach instead interpret the wave function or quantum state as something inherently subjective or about us. For example, one might understand the wave function as representing a kind of information (Fuchs 2003) or knowledge one has about the world, or as saying something about what are rational degrees of belief one should have about, say, where an electron is or the consequences of a subject's actions on a physical system (Fuchs 2011). The fact that such approaches are not just represented but taken seriously by a large part of the physics community raises the issue of whether we as philosophers are licensed to take such a realist attitude to quantum mechanics and engage with the question of what the wave function could be, and what role it has to play in our overall conception of how the world is at its most fundamental independent of us as observers. However, these are not questions that we pursue here. I close with the following.

Since the beginning of quantum mechanics, there has been a disagreement over the correct way to understand and state the changes quantum theory forces us to make to our classical way of thinking about the physical world. For Bohr and Heisenberg, to appreciate the shift required from classical to quantum theory required giving up a fully realist account of physical systems. The quantum state describes only what was or could be observed. Today's antirealists speak more in information-theoretic or Bayesian terms. The alternative account is that beginning with de Broglie and Schrödinger: to appreciate the shift required from classical to quantum theory we must appreciate the wave aspect of both radiation and matter. When we do this, we come to recognize the pervasiveness of entanglement as an objective phenomenon and the necessity of a metaphysics that can make sense of such states—whether that ultimately requires the metaphysics of wave function realism or not is a debate worth having.

6 Postscript on Relativity

In elaborating the main features of quantum mechanics in its various versions, we have mainly ignored the issue of relativity. But, the reader may ask, isn't the only promising candidate for a final theory a relativistic one? So if we are interested in genuine issues of fundamental ontology, shouldn't we be focusing on relativistic versions of quantum mechanics or quantum field theory? There are at least a couple of issues that we would like to address here. The first is about what the relativistic versions of the theories the preceding metaphysical discussions presupposed look like (or whether relativistic versions of these theories even exist). The second is what consequences relativity has for the metaphysical interpretation of these theories.

---

[42] This allows Fuchs 2011 to call those who defend such approaches ψ(read: psi)-ontologists.

Let's start with the first question and the case of Everettian quantum mechanics. Recall that the central idea of Everettian quantum mechanics is to retain the standard quantum formalism. No laws are added describing the behavior of additional hidden variables or collapse mechanisms. When physicists, most notably Dirac, began developing relativistic versions of quantum mechanics, they worked with the standard theory, relativizing the Schrödinger equation, $\hat{H}\psi = -i\hbar\dfrac{\partial\psi}{\partial t}$.[43] So, one might say relativistic quantum mechanics is a version of Everettian quantum mechanics.

Roderich Tumulka, following on an earlier suggestion of Bell, has recently given us a fully relativistic version of the GRW spontaneous collapse theory—at least for the relatively simple case of noninteracting particles (Tumulka 2006). Tumulka bases his account on the version of GRW supplemented with a flash ontology proposed by Bell. Bell himself was quite optimistic about the extension of GRW's theory to a relativistic setting (1987, p. 209). His proposal for the local beables of GRW, the flashes (i.e., collapse events) at space-time locations, are especially amenable to a translation to a relativistic version of the theory. In this version of GRW, the Schrödinger equation is replaced with the Dirac equation, specifying the relativistic evolution of the wave function over time. The wave function is then said to evolve according to the Dirac equation except for when there is a collapse. Tumulka's breakthrough (in particular) was the discovery of a relativistic generalization of the probability per unit time of a GRW collapse— and of the collapse process itself—in both of which the nonrelativistic hypersurfaces of absolute simultaneity are replaced by Lorentz-invariant space-like hyperboloids. The resulting theory approximates the nonrelativistic predictions for the low-velocity case (Tumulka 2006, pp. 839–40).

Successful relativistic versions of Bohmian mechanics have been harder to come by, but constitute active research programs in the physics community. One central problem is that the guidance equation describes the evolution of a configuration of particles over time (Goldstein 2006). If the law describes what happens to a configuration of particles, it is presumably about the simultaneous position states of a system of many particles. It would thus seem to presuppose a notion of absolute simultaneity. But in relativity, there are no absolute facts about what is simultaneous with what; facts about simultaneity are instead dependent on a choice of reference frame. One strategy for the Bohmian that has been pursued is to simply stipulate a preferred frame of reference to support the guidance equation (e.g., Bohm and Hiley 1993). Although any such stipulation will amount to an explicit violation of Lorentz invariance, it will preserve what you might call observational Lorentz invariance: it can easily be shown (for example) that the identity of the preferred frame, in cases like this, will be

---

[43] The first step is to make adjustments in the Hamiltonian, for example, by replacing the kinetic energy element of the Hamiltonian, $\hat{H} = \dfrac{p^2}{2m}$, by $\hat{H} = \sqrt{c^2 p^2 + m^2 c^4}$. See Shankar 1994, p. 564.

undiscoverable by any possible experiment. For further strategies, see the references in Goldstein (2006).

So relativistic versions of quantum mechanics do exist and appear to involve again as in the nonrelativistic case laws describing the evolution of the wave function alone, or the wave function and its impact on the behavior of some additional variables—GRW flashes or Bohmian particles. So it might seem that our interpretive options remain the same as before, assuming a promising relativistic version of Bohmian mechanics can be worked out.

As noted above, however, an important recent article by Wallace and Timpson argues that relativistic considerations should make us question the viability of wave function realism, even if the conceptual objections to the view (expressed in the papers mentioned herein by Allori, Dürr, Goldstein, Tumulka, and Zanghì, Maudlin, and Monton) are ultimately resolvable. This takes us to the second set of issues. A central problem here stems from the fact that in relativity theory, we find that particle number is not conserved. As one popular textbook puts it:

> The union of relativity and quantum mechanics produces the following problem: relativity allows particle production given enough energy.... Consequently the degrees of freedom of a relativistic system are neither fixed nor finite; a system that initially has one particle can evolve into a state with 15 of them. (Shankar 1994, p. 574)

The problem that Wallace and Timpson raise for wave function realism is that if particle number can vary, then the dimensionality of configuration space can vary, and this would appear to undermine the position of the wave function realist who takes configuration space to be the fundamental physical space in which the dynamics play out.

From the fact that in relativistic versions of quantum theory, for example, quantum field theory, particle number is not conserved, Wallace and Timpson argue that:

> no really satisfactory notion of "configuration space" is available for us to formulate wave-function realism.... If there is a "configuration-space" representation of the quantum state, it is given by assigning a (non-normalized) wave-function to each of the infinitely many 3N-dimensional configuration spaces. (2010, p. 707)

Why would no satisfactory notion of configuration space be available if particle number fails to be conserved? The idea seems to be that because configuration space is supposed to be a space with a dimensionality determined by the number of particles in a system, if there is no objective N such that N is the number of particles in the system, then there is no objective fact about the dimensionality of the configuration space. So there would appear to be no objective fact about the nature of the wave function, a or the central object in the wave function realist's ontology, because this wave function is just supposed to be a field over configuration space.

Wallace and Timpson suggest one recourse the wave function realist may take: to say instead that there are many configuration spaces corresponding to each value N may take, and hence many wave functions. But as they point out, this is the wrong solution to the problem, not only because the resulting ontology is unintuitive and unparsimonious. This response doesn't take seriously the reason the problem presented itself to the wave function realist in the first place. Wallace and Timpson continue:

> For particles are not only non-conserved in [quantum field theory], they are non-fundamental: mathematically speaking they are emergent entities supervenient on an underlying field ontology.... This makes particle configurations unattractive—technically as well as conceptually—as the basis for defining the ontology of QFT. (Wallace and Timpson 2010, p. 707)

Wallace and Timpson's proposal is that we reject an interpretation of quantum mechanics that takes it to be a theory about a wave function (a field) on configuration space and instead take it to be a theory about complex and abstract properties of space-time regions. In the rest of this postscript, I address their argument to try to show why the failure of particle number conservation does not undermine wave function realism.

First, note that the wave function realist did not need to consider the shift from nonrelativistic quantum mechanics to quantum field theory to realize that particles are not fundamental. For the wave function realist,[44] all there is fundamentally is a wave function in configuration space. Even if we call it "configuration space" and represent it as $3N$-dimensional where N is the number of particles, the wave function realist will insist that the nature of this space is not grounded in ontologically prior facts about the existence of particles in a three-dimensional space. Rather, this characterization of the wave function is best regarded as an heuristic, a way to connect up the fundamental ontology of quantum mechanics with a microscopic ontology with which we are more familiar, one that we are more easily able to connect up with a macroscopic ontology, one of (nonfundamental) particles. Even if there aren't fundamentally any particles, this will not affect the fact that there is a very high-dimensional space, what is perhaps misleadingly named "configuration space," that the wave function exists on, and that this is the fundamental physical space on which the dynamics play out.

---

[44] For simplicity, let's focus on monistic versions of wave function realism, that is, those focusing on Everettian or collapse versions of quantum theory, as opposed to the dualistic versions that result from the interpretation of Bohmian mechanics and other hidden variables theories.

The wave function realist can allow that the following sentence may be true: The dimensionality of configuration space is $3N$-dimensional, where N is the number of particles in the system. But she should also be clear (and usually she is clear) that this sentence does not express a fact about metaphysical grounding, that is, about how the dimensionality of configuration space is determined. Rather, it expresses a relation between the dimensionality of configuration space and a less fundamental microscopic ontology with which we are more familiar and have an easier of time of connecting to our conception of macroscopic objects.

Even so, there is still the question of how, if particle number is not conserved, there can be any objective, stable fact about the nature of this fundamental space. If, as Shankar notes, particle number can vary from 1 to 15, then what is the dimensionality of configuration space? Is it 3-dimensional or 45-dimensional? I agree with Wallace and Timpson that the wave function realist should not say that there are two configuration spaces, one that is 3-dimensional and one that is 45-dimensional in this case. Rather, the starting point of the wave function realist's response should be that the number of particles does not determine the dimensionality of the configuration space, but the other way around. So it is just important that the wave function in configuration space be such that its behavior over time can ground the existence of however many particles there are. Note that here we are really interested in not the configuration space for small systems like that of 1 or 15 particles, but the configuration space for the system that is the whole universe. This raises the possibility that the configuration space is instead infinite-dimensional and won't need to vary its dimensionality at all to ground the relativistic effects that make for violation of particle number conservation.

An alternative is for the wave function realist to characterize the dimensionality of configuration space altogether differently. If the reason there can be at a certain time N total particles and at a later time M total particles is due to variations in an underlying field ontology, then perhaps it is better to characterize the configuration space primarily in terms of the arrangement of field values. Here is the idea. Don't start with the image of multiple particles spread out in some configuration. Instead, start with the image of a field that can be in different states. Take the dimensionality of the configuration space then to be N where N is the number of possible mutually orthogonal states for this field. N will be infinite. Then each point in the configuration space can be represented by an ordered N-tuple, with each number representing a state of the field. Wallace and Timpson worry about such a proposal that there will be multiple kinds of fields on which one can base such a characterization and as such there will be multiple ways to specify the configuration space and states of the wave function (Wallace and Timpson 2010, p. 708). However, note that the earlier problem is in this way solved. As long as the number of alternative fields is fixed, there will be no fluctuation in the dimensionality of the configuration space. All that is needed is one way to characterize this space such that it can ground the features we observe. If the wave function realist characterizes (though again without grounding) the dimensionality of the configuration space in terms of the states of the field underlying the appearance of particles, not in the number of particles, then the dimensionality of configuration space will remain stable even if the number of particles varies.

In conclusion, there are genuine issues that arise when we consider the transition from nonrelativistic quantum mechanics to relativistic quantum mechanics and quantum field theory. However, even as we shift to a relativistic framework, the central interpretive issues of this volume remain.

References

Albert, David Z. 1992. *Quantum Mechanics and Experience*. Cambridge: Harvard University Press.

Albert, David Z, and Barry Loewer. 1996. Tails of Schrödinger's Cat. In *Perspectives on Quantum Reality: Non-Relativistic, Relativist, and Field-Theoretic*, ed. R. Clifton. Dordrecht: Kluwer, 81–92.

Allori, Valia, Sheldon Goldstein, Roderich Tumulka, and Nino Zanghì. 2007. On the Common Structure of Bohmian Mechanics and the Ghirardi-Rimini-Weber Theory. *British Journal for the Philosophy of Science* 59:353–89.

Allori, Valia, Sheldon Goldstein, Roderich Tumulka, and Nino Zanghì. 2011. Many Worlds and Schrödinger's First Quantum Theory. *British Journal for the Philosophy of Science* 62:1–27.

Barrett, Jeffrey A. 1999. *The Quantum Mechanics of Minds and Worlds*. Oxford: Oxford University Press.

Bassi, A., and Ghirardi G. C. 2003. Dynamical Reduction Models. *Physics Reports* 379:257–426.

Bell, John. 1987. *Speakable and Unspeakable in Quantum Mechanics*. Cambridge: Cambridge University Press.

Belot, Gordon. 2012. Quantum States for Primitive Ontologists. *European Journal for the Philosophy of Science* 2:67–83.

Bohm, David. 1951. *Quantum Theory*. New York: Dover.

Bohm David. 1952. A Suggested Interpretation of the Quantum Theory in Terms of "Hidden" Variables. Reprinted in *Quantum Theory and Measurement*, ed. John Wheeler and Wojciech Zurek. Princeton: Princeton University Press, 1983.

Bohm, David, and Basil Hiley. 1993. *The Undivided Universe: An Ontological Interpretation of Quantum Theory*. London: Routledge.

Bohr, Niels. 1913. On the Constitution of Atoms and Molecules. *Philosophical Magazine*, series 6, 26:1–25.

Bohr, Niels. 1927. The Quantum Postulate and the Recent Development of Atomic Theory. In *The Philosophical Writings of Niels Bohr*, vol. 1. Woodbridge: Ox Bow Press.

Bohr, Niels. 1964. The Structure of the Atom (1922). In *Nobel Lectures, Physics 1942–1962*. Amsterdam: Elsevier, 7–43.

Born, Max. 1964. The Statistical Interpretations of Quantum Mechanics (1954). In *Nobel Lectures, Physics 1942–1962*. Amsterdam: Elsevier, 256–67.

Born, Max, and Albert Einstein. 1971. *The Born-Einstein Letters 1916–1955*. London: Macmillan.

Brown, Harvey. 2010. Reply to "DeBroglie–Bohm Pilot-Wave Theory: Many Worlds in Denial?" In *Many Worlds? Everett, Quantum Theory, and Reality*, ed. Simon Saunders, Jonathan Barrett, Adrian Kent, and David Wallace. Oxford: Oxford University Press.

Brown, Harvey R., and David Wallace. 2005. Solving the Measurement Problem: De Broglie-Bohm Loses Out to Everett. *Foundations of Physics* 35:517–40.

Bub, Jeffrey, and Itamar Pitowsky. 2010. Two Dogmas about Quantum Mechanics. In *Many Worlds? Everett, Quantum Theory, and Reality*, ed. Simon Saunders, Jonathan Barrett, Adrian Kent, and David Wallace. Oxford: Oxford University Press, 433–59.

Cushing, James T. 1994. *Quantum Mechanics: Historical Contingency and the Copenhagen Hegemony*. Chicago: University of Chicago Press.

Deustsch, David. 1985. Quantum Theory as a Universal Physical Theory. *International Journal of Theoretical Physics* 24:1–41.

Dirac, P. A. M. 1930. *The Mathematical Principles of Quantum Mechanics*. Oxford: Clarendon.

Dürr, Detlef, Sheldon Goldstein, and Nino Zanghì. 1992. Quantum Equilibrium and the Origin of Absolute Uncertainty. *Journal of Statistical Physics* 67:843–907.

Einstein, Albert, Boris Podolsky, and Nathan Rosen. 1935. Can Quantum-Mechanical Description of Reality Be Considered Complete? Reprinted in *Quantum Theory and Measurement*, ed. John Wheeler and Wojciech Zurek. Princeton: Princeton University Press, 1983.

Everett, Hugh III. 1957. "Relative State" Formulation of Quantum Mechanics. Reprinted in *Quantum Theory and Measurement*, ed. John Wheeler and Wojciech Zurek. Princeton: Princeton University Press, 1983.

Fine, Arthur. 1986. *The Shaky Game: Einstein, Realism, and the Quantum Theory*. Chicago: University of Chicago Press.

Fuchs, Christopher. 2003. Quantum Mechanics as Quantum Information, Mostly. *Journal of Modern Optics* 50:987–1023.

Fuchs, Christopher. 2011. QBism: The Perimeter of Quantum Bayesianism. Available online at http://arxiv.org/abs/1003.5209.

Ghirardi, GianCarlo, Alberto Rimini, and Tullio Weber. 1986. Unified Dynamics for Microscopic and Macroscopic Systems. *Physical Review D* 34:470–71.

Goldstein, Sheldon. 2006. Bohmian Mechanics. *Stanford Encyclopedia of Philosophy*. Available online at http://plato.stanford.edu.

Greaves, Hilary. 2007. Probability in the Everett Interpretation. *Philosophy Compass* 2:109–28.

Hawthorne, John. 2010. A Metaphysician Looks at the Everett Interpretation. In *Many Worlds? Everett, Quantum Theory, and Reality*, ed. Simon Saunders, Jonathan Barrett, Adrian Kent, and David Wallace. Oxford: Oxford University Press, 144–54.

Healey, Richard. 1989. *The Philosophy of Quantum Mechanics: An Interactive Interpretation*. Cambridge: Cambridge University Press.

Heisenberg, Werner. 1927. The Physical Content of Quantum Kinematics and Mechanics. Reprinted in *Quantum Theory and Measurement*, ed. John Wheeler and Wojciech Zurek. Princeton: Princeton University Press, 1983.

Heisenberg, Werner. 1930. *The Physical Principles of the Quantum Theory*. Chicago: University of Chicago Press.

Hughes, R. I. G. 1989. *The Structure and Interpretation of Quantum Mechanics*. Cambridge, Mass.: Harvard University Press.

Ismael, Jenann. 2009. Quantum Mechanics. *Stanford Encyclopedia of Philosophy*. Available online at http://plato.stanford.edu.

Jammer, Max. 1966. *The Conceptual Development of Quantum Mechanics*. New York: McGraw-Hill.

Kemble, Edwin. 1958. *The Fundamental Principles of Quantum Mechanics, with Elementary Applications*. New York: Dover.

Kumar, Manjit. 2008. *Quantum: Einstein, Bohr, and the Great Debate about the Nature of Reality*. New York: Norton.

Lewis, Peter. 2004. Life in Configuration Space. *British Journal for the Philosophy of Science* 55:713–29.

Maudlin, Tim. 2007. Completeness, Supervenience, and Ontology. *Journal of Physics A* 40: 3151–71.

Maudlin, Tim. 1994. *Quantum Non-Locality and Relativity*. Oxford: Blackwell.

Mermin, David. 1981. Quantum Mysteries for Anyone. *Journal of Philosophy* 78:397–408.

Monton, Bradley. 2006. Quantum Mechanics and 3N-dimensional Space. *Philosophy of Science* 73:778–89.

Przibram, K. 1967. *Letters on Wave Mechanics*. New York: Philosophical Library.

Saunders, Simon. 1993. Decoherence, Relative States, and Evolutionary Adaptation. *Foundations of Physics* 23:1553–85.

Schrödinger, Erwin. 1935. The Present Situation in Quantum Mechanics. Reprinted in *Quantum Theory and Measurement*, ed. John Wheeler and Wojciech Zurek. Princeton: Princeton University Press, 1983.

Shankar, R. 1994. *Principles of Quantum Mechanics*, 2nd ed. New York: Springer.

Spekkens, Robert W. 2007. Evidence for the Epistemic View of Quantum States: A Toy Theory. *Physical Review A* 75:032110.

Tumulka, Roderich. 2006. A Relativistic Version of the Ghirardi-Rimini-Weber Model. *Journal of Statistical Physics* 125:821–40.

Valentini, Antony. 2010. De Broglie–Bohm Pilot-Wave Theory: Many Worlds in Denial? in *Many Worlds? Everett, Quantum Theory, and Reality*, ed. Simon Saunders, Jonathan Barrett, Adrian Kent, and David Wallace. Oxford: Oxford University Press

Van Fraassen, Bas. 1991. *Quantum Mechanics*. Oxford: Clarendon Press.

Von Neumann, John. 1932/1996. *The Mathematic Foundations of Quantum Mechanics*. Princeton: Princeton University Press.

Wallace, David. 2010. Decoherence and Ontology. In *Many Worlds? Everett, Quantum Theory, and Reality*, ed. Simon Saunders, Jonathan Barrett, Adrian Kent, and David Wallace. Oxford: Oxford University Press.

Wallace, David. 2003. Everett and Structure. *Studies in the History and Philosophy of Modern Physics* 34:87–105.

Wallace, David, and Christopher Timpson. 2010. Quantum Mechanics on Spacetime I: Spacetime State Realism. *British Journal for the Philosophy of Science* 61:697–727.

Wigner, Eugene. 1961. Further Remarks on the Mind-Body Problem. Reprinted in *Quantum Theory and Measurement*, ed. John Wheeler and Wojciech Zurek. Princeton: Princeton University Press, 1983.

1

Wave Function Realism

DAVID Z ALBERT

A hundred years ago, physics aspired to produce a complete, and unified, and seamless, and philosophically realistic account of the entirety of nature. It aspired to tell us straightforwardly *what the world is*. It aspired, that is, to settle questions of metaphysics.

And all of that came to look somehow quaint and childish under the spectacular assault of quantum mechanics. Bohr and his circle quickly became convinced that the so-called problem of measurement was no ordinary scientific problem; that, as a matter of fact, it was like nothing we had ever encountered before; that what we were up against here were the ultimate limits of the scientific project itself. The idea was that the business of representing the world in scientific terms had now been discovered to depend on one first establishing some sort of a *boundary* between observed and observer, or between mind and body, or between subject and object, or between what could be spoken of in classical language and what could not. Moreover, Bohr and his circle were convinced that this boundary itself—even though it was something on which our most fundamental scientific ideas crucially depended—was not the sort of thing on which scientific investigation could ever *shed any light*: that there was something permanently mysterious about it, or something arbitrary, or something subjective, or something merely verbal. A host of what had seemed like perfectly sensible questions about the physical situation of the world were declared somehow unaskable or (worse) unintelligible. There was talk of the "*renunciation*, in each experimental arrangement, of one or the other of the aspects of the description of physical phenomena." There was talk of the necessity of some "radical revision of our attitude towards physical reality."

This strikes most of us nowadays as weird, glib, scary, oppressive, intolerant stuff. But it fell on the physics of that other time like a divine pronouncement and shut down the conversation about these matters, and it has only been over the past 25 years or so that this dark fog has finally begun to lift, and that the puzzle about measurement in quantum mechanics has at long last been transformed into an unambiguously scientific problem, like other unambiguously scientific problems, and that physics is again in a position at least to *aspire* to tell us what the world straightforwardly is.

Now, it has always been part and parcel of all of the old, creepy, antirealist ways of thinking about quantum mechanics that one looked at quantum mechanical wave functions not as representing physical situations *directly* but as representing, say, *what observers know* of such situations, or as representing imaginary *ensembles* of such situations, or as representing *the probabilities of the outcomes of measurements* in such situations, or something like that.

And it has always been very much of the essence of digging one's way out of that sort of confusion—that is, it has always been very much of the essence of the project of quantum mechanical *realism* (in whatever particular form that realism takes—Bohm's theory, or modal theories, or Everettish theories, or theories of spontaneous localization), that one learn to think of wave functions differently. And this chapter is going to be taken up with what seems to me to be the simplest and, most straightforward, and most flat-footed way of thinking— in this new realistic spirit—about quantum mechanical wave functions, which is to think of them as *concrete physical objects*.

The most striking and controversial feature of this approach is undoubtedly that the stage on which such objects must make their appearance, the stage (that is) on which any such understanding of quantum mechanics is going to have to depict the history of the world as playing itself out, is a mind-numbingly high-dimensional space—a space (more particularly) that is isomorphic to what is referred to in more conventional understandings of quantum mechanics as the *configuration space* of the world, a space whose dimensionality is three times as large as the total number of elementary particles in the universe. And it is obviously going to be incumbent on any such understanding to explain our vivid and deep-seated impression to the contrary. And I will attend to that in a minute.

But it will be best, first, to sketch out the basic metaphysics. The sorts of physical objects that wave functions are, on this way of thinking, are (plainly) *fields*—which is to say that they are the sorts of objects whose states one specifies by specifying the values of some set of numbers at every point in the arena in which they live, the sorts of objects whose states one specifies (in this case) by specifying the values of two numbers (one of which is usually referred to as an *amplitude*, and the other as a *phase*) at every point in the configuration space of the universe.

The values of the amplitude and the phase are thought of (as with all fields) as intrinsic properties of the points in the high-dimensional space with which they are associated. So, for example, the fact that the integral over the entirety of the configuration of the square of the amplitude of the universe's wave function is invariably equal to one is going to have to be thought of not as following analytically from the sorts of physical objects wave functions *are* (which it certainly cannot) but as a *physical law*, or perhaps as an *initial condition*.

What physical role this object plays in the world, precisely, will depend on precisely how the measurement problem gets solved.

On Bohm's theory, for example, the world will consist of exactly two physical objects. One of those is the universal wave function, and the other is the universal *particle*. And the story of the world consists, in its entirety, of a continuous succession of changes of the *shape* of the former and a continuous succession of changes in the *position* of the latter. And the dynamical laws that *govern* all those changes—that is, the Schrödinger equation and the Bohmian guidance condition—are completely deterministic, and (in the high-dimensional space in which these objects live) completely *local*.

On the Ghirardi-Rimini-Weber (GRW) theory (or, for that matter, on *any* theory of collapse), the world will consist of exactly one physical object—the universal wave function. What happens, *all* that happens, is that the function changes its shape in accord with the theory's dynamical laws. And those changes are not entirely continuous, and the laws governing them are not entirely deterministic and (even in the high-dimensional configuration space) not entirely local.

The particularly urgent question (again) is where, in this picture, all the tables, and chairs, and buildings, and people are. The particularly urgent question is how it can possibly have come to pass, on a picture like this one, that there appear to us to be *multiple* particles moving around in a *three-dimensional* space.

And the thing to keep in mind is that what it is to be a table or a chair or a building or a person is—at the end of the day—to occupy a certain location in the causal map of the world. The thing to keep in mind is that the production of geometrical appearances is—at the end of the day—a matter of *dynamics*.

Think (to begin with) of a real, concrete, D-dimensional space, with a single classical particle floating around in it, under the influence of a classical Hamiltonian H. And suppose that there is some orthogonal coordinatization of this D-dimensional space—call it C—on which H happens to take the form:

$$H = \Sigma_i m_i((dx_{(3i-2)}/dt)^2 + (dx_{(3i-1)}/dt)^2 + (dx_{(3i)}/dt)^2)$$

$$+ \Sigma_{i \neq j} V_{ij}((x_{(3i-2)} - x_{(3j-2)})^2 + (x_{(3i-1)} - x_{(3j-1)})^2 + (x_{(3i)} - x_{(3j)})^2), \qquad (1.1)$$

where i and j range over the integers from 1 to $D/3$ inclusive.[1] Looked at in C, then, the position coordinates of this particle will evolve in time exactly as if they were the coordinates of $D/3$ classical particles floating around in a three-dimensional space, and *interacting* with one another in accord with a law that is built up out of the *geometrical structures* of that three-dimensional space and which depends

---

[1] Of course, if there is one such coordinatization, then there will necessarily be an infinite number, each of which is related to C by means of some combination of three-dimensional translations and rotations and boosts.

on the interparticle *distance* in that three-dimensional space, which is invariant under the *symmetries* of that three-dimensional space and which has the familiar mathematical form:

$$H = \Sigma_i m_i((dx_i/dt)^2 + (dy_i/dt)^2 + (dz_i/dt)^2) + \Sigma_{i \neq j} V_{ij}((x_i - x_j)^2$$

$$+ (y_i - y_j)^2 + (z_i - z_j)^2). \tag{1.2}$$

This particle, in this space, moving around under the influence of the Hamiltonian in equation 1.1, *formally enacts* (you might say) a system of $D/3$ classical three-dimensional particles—the ith of which is the projection of the world particle onto the $(3i\text{-}2, 3i\text{-}1, 3i)_C$ subspace of the D-dimensional space in which the world particle floats.

If we pretend (for a moment) that the laws of ordinary three-dimensional Newtonian mechanics, together with the familiar three-dimensional Hamiltonian in equation 1.2, can accommodate the existence of the tables and chairs and baseballs of our everyday experience of the world,[2] then we shall be able to speak (as well) of formal enactments of tables and chairs and baseballs, by which we will mean the projections of the position of the world particle onto tensor products of various of the $(3i\text{-}2, 3i\text{-}1, 3i)_C$ subspaces of the D-dimensional space in which the world particle floats.[3] And these formally enacted tables and chairs and baseballs are clearly going to have precisely the same causal relations to one another, and to their constituent formally enacted particles, as genuine tables and chairs and baseballs and their constituent particles do.

And in so far (then) as we have anything in the neighborhood of a functionalist understanding of what it is to be a table or a chair or a baseball—in so far (that is) as what it is to be table or a chair or a baseball can be captured in terms

---

[2] Of course, it isn't true that the laws of ordinary three-dimensional Newtonian mechanics, together with a Hamiltonian like the one in equation 1.2, can accommodate the existence of the tables and chairs and baseballs of our everyday experience of the world. Those laws, after all, can't even account for the stability of individual atoms, much less the tendency of such atoms to cohere into stable macroscopic objects. That, among other reasons, is why we need quantum mechanics. All of that is beside the point. The question we want to focus on here is (as it were) whether it is any harder for there to be tables and chairs and baseballs in a $3N$-dimensional world consisting of a single material point than it is for there to be tables and chairs and baseballs in a three-dimensional world consisting of N classical particles. The question (more precisely) is this: Supposing that there could be tables and chairs and baseballs in a three-dimensional world consisting of N classical particles moving around under the influence of a Hamiltonian like the one in equation 1.2—whatever, exactly, it might mean to suppose such a thing—is there then anything that stands in the way of there being tables and chairs and baseballs in a $3N$-dimensional world consisting of a single material point moving around under the influence of a Hamiltonian like the one in equation 1.1?

[3] It would be more precise, I suppose, to speak not of the formal enactment of this or that table or chair or particle but of the formal enactment of this or that total three-dimensional physical situation *involving* a table or a chair or a particle. The former, easier, more efficient way of speaking will serve well enough, I think, as long as we keep its more accurate expansion in the backs of our minds.

of the causal relations of these objects to one another and to their constituent particles and so on—then these formally enacted tables and chairs and baseballs and particles must really *be* tables and chairs and baseballs and particles. And in so far as what it is to be a *sentient observer* can be captured in terms like these, then projections of the world particle onto those particular tensor products of three-dimensional subspaces of the D-dimensional space that correspond to such "observers" are necessarily going to have psychological experience. It is plainly going to *appear* to such observers that the world is three-dimensional!

Of course, in so far as we confine our considerations to the case of classical mechanics, all of this is a fairly idle academic entertainment—because in the classical case there is no reason to take these high-dimensional pictures seriously, because in the classical case there is always already an option of saving the three-dimensional appearances by means of an exact and universal and fundamental theory of a thoroughly three-dimensional world. But the point of all this is, of course, that the quantum mechanical case is different—the point of all this is that in the quantum mechanical case, the reasons for taking these high-dimensional pictures seriously are, on the face of it, very powerful.[4] And the thought is that the classical entertainment we have just been through has in it the crude beginnings of an account—or, rather, of a set of accounts—of how it happens that there can be tables and chairs and baseballs and people and three-dimensional appearances in Bohm's theory, and in the GRW theory, and in any other workable solution to the measurement problem in which high-dimensional quantum mechanical wave functions play an essential and ineliminable role.

The business of actually filling in the details of these accounts is not an altogether trivial matter and needs to be approached separately, and anew, for each particular way of solving the measurement problem, and requires that we attend carefully to exactly how it is that the things we call particles actually manifest themselves in our empirical experience of the world.

In the Bohmian case, for example, there is patently *not* going to be any coordinate system of the high-dimensional space in which the coordinates of the world particle evolve in time exactly as if they were the coordinates of $D/3$ classical particles floating around in a three-dimensional space and of course, we wouldn't want there to be, because the world we are looking to enact in this case is not a classical one but a quantum mechanical one. But what it is to enact a system of quantum mechanical particles is (on the other hand) not, on the face of it, all that easy to say. The exact trajectories of quantum mechanical

---

[4] Whether these reasons are in fact as powerful as I take them to be is a topic of lively debate, both within this volume and elsewhere—and I have a good deal to say about that in a forthcoming book. What seems most useful here (on the other hand) is merely to present as clear, simple, and concise a picture of the positive view as I can—without going into the details of any of the arguments, pro or con—to provide a convenient and agreed-on target to shoot at.

particles—if, indeed, quantum mechanical particles *have* any exact trajectories—certainly do not count among those features of the world to which we can ever have any direct observational access. And so there can be no exact and particular claims about the motions of particles—over and above what we have from the Schrödinger equation and the Born rule—that we ought properly to be in the business of trying to underwrite in this case.

And the situation in the GRW theory is apparently even less auspicious. In that latter theory (after all)—at least on the fundamental level—there is nothing particulate whatsoever.

But all of this, as I will argue in another place, can be sorted out and there turn out to be serviceable accounts of the emergence of everyday appearances in these theories which do indeed, at the end of the day, run very much along the lines of the classical entertainment we just went through.

2

Primitive Ontology and the Structure of Fundamental Physical Theories

VALIA ALLORI

For a long time, it was believed that it was impossible to be realist about quantum mechanics. It took quite a while for the researchers in the foundations of physics, beginning with John Stuart Bell (1987), to convince others that such an alleged impossibility had no foundation. These days there are several quantum theories that can be interpreted in a realist fashion, among which Bohmian mechanics, the Ghirardi-Rimini-Weber (GRW) theory, and the many-worlds theory.

The debate, though, is far from being over. In what respect should we be realist regarding these theories? At least two different proposals have been made: on one hand, some insist on a direct ontological interpretation of the wave function as representing physical bodies; on the other hand are those who claim that quantum mechanics is not really about the wave function.

In this chapter I present and discuss one proposal of the latter kind that focuses on the notion of *primitive ontology*.

1 Wave Function Ontology

There is a realist take on quantum theories according to which quantum mechanics is a theory about the behavior of an object called the wave function.[1] That is, the wave function mathematically represents a real, physical field that constitutes physical objects. For this reason, such a view has been called *wave function ontology*. One of the strongest arguments for this view is an argument by analogy. If in a physical theory there is a fundamental equation for the evolution of a given mathematical object, generally we feel justified to take this entity to represent physical objects. Consider classical mechanics: the fundamental equation of this theory is Newton's equation that describes the temporal evolution of a point

---

[1] See Albert 1992, 1996; Albert and Loewer 1996; Lewis 2004; Lewis 2005, Lewis 2006; Wallace 2002; 2003, and references therein.

in three-dimensional space. It is natural to interpret such object as describing a particle whose dimensions are negligible, and this is exactly the way we take it: we conclude that classical mechanics is a theory that describes the behavior of point-particles. By analogy, we should do the same in quantum mechanics: given that in this theory there is a fundamental equation, Schrödinger's equation, for the temporal evolution of the wave function, we are entitled to take the wave function to represent physical objects as well.

As a consequence of this view, physical space is not the traditional three dimensional space. Rather, it is the space on which the wave function is defined: this is called "configuration space." Historically, configuration space has been introduced in classical mechanics for mathematical purposes. It is constructed from three-dimensional physical space: if there are N point-particles, each with position r_i in three-dimensional space, then configuration space is defined as the space of the configurations of all particles. That is, an element q of configuration space is given by $q = (r_1, r_2, \ldots, r_N)$. As a consequence, if there are N particles in the universe, configuration space has dimension $M = 3N$. Observe that if one maintains that physical bodies are represented by the wave function, then literally there are no particles, and therefore there is no real reason to call such space configuration space. The proponents of this view realize this, but the name sticks nonetheless.

Because the proposal is to take the wave function to represent physical objects, it seems natural to take configuration space as the *true* physical space. But clearly, we do not seem to live in configuration space. Rather, it seems obvious to us that we live in three dimensions. Therefore, a proponent of this view has to provide an account of why it seems *as if* we live in a three-dimensional space even though we do not. Connected to that problem, we should explain how to "recover the appearances" of macroscopic objects in terms of the wave function. Using Wilfrid Sellars's terminology (Sellars 1962), we need to reconcile the *scientific image* (the image of the world that our best scientific theories are giving us) with the *manifest image* (the image of the world that we ordinarily experience). This is something that proponents of this view are working on. Whether this project is succeeding and whether it is in principle possible have been challenged elsewhere.[2]

2 Primitive Ontology

There are people who find the view just presented unsatisfactory and put forward different alternatives. In this chapter I focus on the proposal that involves the notion of primitive ontology. Other positions that do not explicitly refer to such a notion have been proposed,[3] but they are not discussed here.

---

[2] See Monton 2002, 2006; and Allori forthcoming.
[3] See Monton 2002, 2006; Maudlin 2007a.

The notion of primitive ontology was first proposed in Dürr et al. (1992) and Goldstein (1998), and then discussed in a little more detail in Allori et al. (2008). The main idea is that all fundamental physical theories, from classical mechanics to quantum theories, share the following common structure:

- Any fundamental physical theory is supposed to account for the world around us (the manifest image), which appears to be constituted by three-dimensional macroscopic objects with definite properties.
- To accomplish that, the theory will be about a given *primitive* ontology: entities living in three-dimensional space or in space-time. They are the fundamental building blocks of everything else, and their histories through time provide a picture of the world according to the theory (the scientific image).
- The formalism of the theory contains primitive variables to describe the primitive ontology, and nonprimitive variables necessary to mathematically implement how the primitive variables will evolve in time.[4]
- Once these ingredients are provided, all the properties of macroscopic objects of our everyday life follow from a clear explanatory scheme in terms of the primitive ontology.

Several questions come to mind at this point—from clarifications about the notion of primitive ontology to questions about the motivations to endorse this view instead of the wave function ontology. I do not directly compare the two approaches here, because this has been done elsewhere (Allori forthcoming). Rather, I present the primitive ontology idea and its framework per se, in particular analyzing the many roles (metaphysical, epistemological, and explanatory) that the primitive ontology has in the scientific enterprise. In doing so, some other questions will receive an answer:

- What is the primitive ontology for and why is there an emphasis on its three-dimensionality (section 3)?
- Is there a rule to identify the primitive ontology of a theory (section 4)?
- What are the differences between the primitive and the nonprimitive variables (section 5)?
- What is the connection between the primitive ontology and the explanatory power of a theory (section 6)?
- How can we apply the primitive ontology framework to quantum theories (section 7)?
- What is the status of the nonprimitve variables (in particular wave function) in this picture (section 8)?
- What is the connection between symmetry properties of a theory and its primitive ontology (section 9)?

---

[4] The metaphysical status of such nonprimitive variables is up for debate (see section 8, for instance), but surely they do not represent physical objects.

3 Scientific and Manifest Image

The primitive ontology proposal is tightly connected to a particular understanding of what physics is, what it does, and how it does it. In other words, it is connected to a particular understanding of what the scientific image is, how we arrive at it, and how it relates to the manifest image. The starting idea is that when a scientist proposes a fundamental physical theory, she already has in mind what the theory is fundamentally about: the primitive ontology.[5] This is the *metaphysical* role of the primitive ontology: it tells us what the world is made of according to the theory.

But how does the scientist choose the primitive ontology? How do we go from the manifest to the scientific image? Or, how do we change from one old scientific image to a new one? To simplify and cut a long story short, in the words of Albert Einstein (1936), "the whole of science is nothing more than a refinement of everyday thinking."The scientific image typically starts close to the manifest image, gradually departing from it if not successful to adequately reproduce the experimental findings. The scientific image is not necessarily close to the manifest image, because with gradual departure after gradual departure we can get pretty far away. In fact, historically we went from the manifest image of a table being continuous and solid to the Newtonian scientific image of the table being composed of microscopic particles and mostly empty. The point, though, is that the scientist will typically tend to make minimal and not very radical changes to a previously accepted theoretical framework. First, she might attempt to keep the same primitive ontology as the old theory, perhaps changing the law with which it evolves. If that fails, she might go for a different primitive ontology that still will not radically change her ways of understanding things. In other circumstances, she might move to a theory that will provide a better explanation.

The situation is complex, and a lengthier discussion should perhaps be required, but let me clarify the main idea with a concrete example. At some point scientists attempted to explain thermal phenomena, such as two bodies in contact reaching the same temperature, positing a primitive ontology not too distant from the manifest image: heat was postulated to be a thermal fluid that passed from one body to the next. When this hypothesis did not work, a less "manifest" primitive ontology was proposed: each body was considered to be composed of many microscopic particles, and heat was understood as the motion of such particles. This is essentially Boltzmann's approach to thermodynamics, in which thermal phenomena are recovered from a Newtonian picture of the world when there is a very large number of particles.[6] The particle primitive ontology was arguably suggested by the experimental failure of the more

---

[5] The qualification "fundamentally" is connected with the role of the primitive ontology in the structure of the theory, as we will see in section 6.

[6] For a discussion, see Goldstein 2001a and Albert 2000.

straightforward choice, the thermal fluid primitive ontology. At the same time, though, a primitive ontology of particles was not a radical choice, given that it was the primitive ontology of Newtonian mechanics, an already successful theory in other domains. Hence, roughly, in this sense the scientific image departed gradually from the manifest image to cope with the experimental results.

Through history, the primitive ontologies of the various fundamental physical theories have changed a lot: people have considered the world as made of fields, particles, flashes,[7] strings, and so on. These proposed primitive ontologies have something in common: they all are in three-dimensional space, or in space-time.[8] Why is that? Because although it seems reasonable that we might be mistaken about what kind of entities the world is made of (fields, particles, and so on), to give up the idea that matter lives in three-dimensional space and evolve in time seems too much, especially if there is no need for it. A primitive ontology in the familiar three-dimensional space evolving in time (or a space-time primitive ontology) is the natural metaphysical choice, if the theory with such a primitive ontology can be empirically and explanatory adequate (namely, a good compromise between getting the empirical predictions right and providing a satisfactory explanation of the phenomena). As in classical mechanics, it seems most convenient to explain, if possible, the behavior of familiar macroscopic bodies postulating that they are composed of microscopic entities in three-dimensional space that constitute the fundamental building blocks of everything else. In fact, as we will see later, we can employ a clear scheme, developed in the framework of classical theories, to explain the properties of three-dimensional macroscopic objects in terms of the properties of their three-dimensional microscopic constituents. We will also see how the same primitive ontology framework can be extended to quantum theories. Because of this, also in quantum theories we can account, at least in principle, for the macroscopic world along the lines of classical mechanics.

The primitive ontology provides us with a clear metaphysical picture of the world. So does the wave function ontology: the world is made of *stuff* represented by the wave function. One difference between the approaches is that the primitive ontology is in three-dimensional space (or in space-time), whereas the wave function is not. As a consequence of this, in the case of the wave function ontology, the scientific image does not have much in common with the previously accepted Newtonian picture. This is not true in the case of theories with a primitive ontology. In contrast to the case of wave function ontology, the primitive ontology approach reflects the desire to keep the scientific image closer to the classical way of understanding things, given that it is possible. The reason for this attitude, as we just saw, is obvious: if you can account for everything that

---

[7] See section 7.

[8] Even if the space of string theory is 10-dimensional, because all dimensions but three are wrapped up on themselves ("compactified"), physical space is, for all practical purposes, three-dimensional.

you need to account for employing already successful and well-tested explanatory techniques, why not do so?

Why the qualification "primitive ontology," instead of just "ontology" simpliciter? First, the idea is that the primitive ontology does not exhaust all the ontology—it just accounts for physical objects. Other things might exist (numbers, mathematical objects, abstract entities, laws of nature, and so on), and some of them (like natural laws) might be described by other objects in the ontology of a fundamental physical theory. We will see in section 8 how this could be true for the wave function. For more on the qualification "primitive," see section 6. Now we turn to the general structure of fundamental physical theories in the primitive ontology framework.

4 The Structure of Fundamental Physical Theories

Assume that the idea of scientific image just discussed is correct. Hence, the scientist formulating a given theory will make a metaphysical hypothesis and develop her theory around it. Physics works through mathematics: a theory contains several mathematical objects, some with a physical significance, others without. The point here is that *this is established once the theory is proposed*: there is already a natural interpretation for each mathematical object, namely, the one the proponent of the theory intended to give them! The scientist's choice of what physically exists in the world will more or less automatically determine the mathematical object to represent it. A fundamental physical theory aims to describe not only what physical bodies there are but also how they evolve in time. Because of this, in addition to the variables describing the primitive ontology, the theory also contains some other equations, whose solutions describe how the primitive ontology moves through space in time.

The mathematical formalism of a theory therefore has a *history* that constrains the interpretation of its formalism: the theory started with a metaphysical position and its appropriate mathematical representation, and it continued with the implementation of the suitable mathematical apparatus necessary to determine how the primitive ontology evolves. For this reason, the argument by analogy already discussed for the wave function ontology view is misguided: it assumes in fact that the mathematical formalism of a theory can be interpreted a posteriori, whereas it was fixed a priori by the physicist when she formulated the theory. Therefore, there is *no* rule to determine the primitive ontology of a theory. Instead, it is a matter of understanding how the theory was introduced, how it has developed, and how its explanatory scheme works (for this, see section 6). Once the scientist sets up the theory, the metaphysical picture it provides has already been defined, and there is very limited freedom of reinterpreting the formalism (at least with the limitations exemplified by the case of classical electrodynamics, as we will see shortly).

Let us explicitly see how this framework works with the aid of some examples. As anticipated, a clear case of a fundamental physical theory with a primitive ontology is classical mechanics. In this theory, physical objects are taken to be particles, and Newton's equation captures the temporal evolution of these objects viathe introduction of forces and masses. These are "additional" variables in classical mechanics in the sense that they were added into the theory to account for the behavior of the primitive ontology.

Another example is given by classical electrodynamics (CED). The theory was developed initially from classical mechanics to account for the evolution of charged particles. New mathematical entities were introduced—the electromagnetic fields. Are the fields a part of the primitive ontology of this theory? Actually, we can have different answers. On one hand, we can insist that the fields were added in the theory to account for the experimental trajectories of charged particles. If so, we have a theory, call it CEDp, in which fields do not represent matter, which is made only of particles. On the other hand, we might be inclined to think that the previous primitive ontology of particles was incomplete, and the fields indeed represent something in the material world. In this case we have a different theory, CEDpf, with a primitive ontology of particles *and* fields. In other words, the variables describing the electromagnetic fields can be regarded as nonprimitive, in the sense that their role in the theory is not to describe physical bodies but to implement the empirically correct behavior of physical bodies (Dürr et al. 1992). Instead, other considerations (such as the fact that the electromagnetic fields have their own law of temporal evolution and there is energy associated with them) have led others to think that the electromagnetic fields represent part of the basic furniture of physical reality (see, for instance, Maudlin 2007a).

Be that as it may, to sum up, all these theories have a dual structure: the primitive variables that specify what matter is, and some other variables that determine its temporal development (its dynamics).

5 Primitive and Nonprimitive Variables

The histories of the primitive ontology—that is, their evolution in space through time—provide the metaphysical picture of the world, and they are produced with the aid of (some of the) nonprimitive variables. Just like a computer program generates an *output*, the fundamental physical theory "generates" the histories of the primitive ontology. As the computer program needs certain internal variables to produce its output, the theory needs additional variables to implement the law of motion for the primitive ontology. Note that we could use different internal variables to obtain the same histories for the primitive ontology. If we do so, we still have fundamentally the very same theory: two theories with the same histories of the primitive ontology can be regarded as *physically* equivalent because they provide us with the very same picture of the world.

This notion of physical equivalence between theories was introduced in Allori et al. (2008) in the framework of quantum mechanics. Nonetheless, it is not necessary to go to quantum theories to give an example of physically equivalent theories. Here is a very simple example of physical equivalent theories. If a force is conservative, it can be defined as the opposite of the gradient of the potential. This particular mathematical operation involves derivatives, and because of this, it is always possible to find two different potentials that give rise to the same histories of the primitive ontology: any two potentials that differ by a constant will do the trick. In fact, they both give rise to the same force (and therefore the same histories of the primitive ontology), given that the derivative of any constant is always zero. Hence, two theories with such potentials will be physically equivalent.

To conclude, two different theories with the same histories of the primitive ontology, no matter how they are implemented, describe the same physical world. The rest is details; how the dynamics for the primitive ontology is implemented is not important in this regard. This stresses the *epistemological* role of the primitive ontology: we only need to know its histories to recover the empirical data, given that the same histories could have been produced by different mathematical variables, as the previous example showed.

6 The Explanatory Scheme of Fundamental Physical Theories

A microscopic primitive ontology grounds a scheme of explanation that allows one to determine the properties of macroscopic physical objects in terms of the behavior of the primitive ontology. In fact, in classical mechanics any physical body (gases, fluids, and solids) is satisfactorily described as a collection of particles. The story the theory tells us about the macroscopic world is a "geometrical story"—a table is just a table-shaped cluster of microscopic primitive ontology. Once the primitive ontology and its temporal evolution are given, everything else follows: the solidity of a table, the localization of a comet, the transparency of a pair of glasses, the liquidity of the water in a bottle, the compressibility of the air in a room, and so on. Arguably, in classical mechanics (as well as in classical electrodynamics) we can identify macroscopic properties more or less straightforwardly given how the microscopic primitive ontology combines and interacts to form complex bodies.[9]

Let us see how that works by way of some examples. First, we can explain why a table is solid on the basis of the fact that it is composed of particles that interact electromagnetically such that it is impossible for another object (for instance, my hand) to penetrate them. Next, suppose we want to account for the

---

[9] An antireductionist would object to this, but granting that reductionism is possible, this is how it is supposed to work.

fact that a comet has a given localization at a given time. One can accomplish this in terms of the microscopic components of the comet and their interaction with each other: the particles interact to form a solid object whose motion (and therefore its localization at different temporal instants) can be just as effectively described by its center of mass. Also, the transparency of an object such as a pair of glasses can be explained in terms of the electromagnetic forces acting between the particles composing the glasses, which are such that incoming light rays will pass through them. Similarly for fluids: a property like the liquidity of water can be explained in terms of the very weak interaction between the microscopic constituents of water that allow it to take the shape of its container. In addition, the behavior of gases is accounted for by considering them as composed by noninteracting particles colliding with one another. This is what happens when we derive thermodynamics from statistical mechanics: what in thermodynamics we call pressure, volume, temperature of a gas are derived from the fact that gases are made up of moving particles. Given that air is a gas, and a gas is just a collection of noninteracting particles, we can also explain why air is compressible: it is possible to reduce the distance between the particles almost as much as we want.

These examples show how we have a clear and straightforward scheme of explanation in the classical framework: given the primitive ontology at the microscopic level, one can employ standard methods to determine the properties of familiar macroscopic objects. This is possible because classical theories have a primitive ontology, so for any other fundamental physical theory with a primitive ontology we could employ an explanatory scheme developed along the same lines.

Thus, in this sense the primitive ontology is the most fundamental ingredient of the theory. It grounds the "architecture" of the theory: first we describe matter through the primitive variables, then we describe its dynamics, implemented by some nonprimitive variables, and that's it. All the macroscopic properties are recoverable. This summarizes the *explanatory* role of the primitive ontology. This is also connected with the "primitiveness" of the primitive ontology: even if the primitive ontology does not exhaust all the ontology, it makes direct contact between the manifest and the scientific image. Because the primitive ontology describes matter *in the theory* (the scientific image), we can directly compare its macroscopic behavior to the behavior of matter *in the world of our everyday experience* (the manifest image). Not so for the other nonprimitive variables, which can only be compared indirectly in terms of the ways they affect the behavior of the primitive ontology.

7   Quantum Mechanics with Primitive Ontology

Classical mechanics and classical electrodynamics provide two paradigmatic examples of how physics tells us about the world: in the scientific image there are the primitive variables that describe matter microscopically, and the manifest image, in which there are macroscopic objects with their properties, is obtained considering the histories of the primitive ontology in the appropriate macroscopic limit. It is a very nice explanatory scheme, straightforward and clear. Too bad it seems we have to abandon it once we consider quantum mechanics. In fact, several extremely strong assertions have been made about quantum theories—from the claim it is impossible to be realist if quantum mechanics is true, to the idea that the act of observation can affect reality, to the insistence that the "old," classical way of understanding the world we just described is no longer suitable.

The reasons for these attitudes can be perhaps understood by briefly recalling the history of quantum mechanics. At the end of the nineteenth century, the Newtonian picture of the world was commonly accepted, even if there were several puzzles: there were experiments whose results did not come out as the theory predicted. Some of them suggested the idea of *quantization*, a discretization of the values certain physical quantities can assume that does not substantially challenge the classical hypothesis that physical objects are made of particles. Other results suggested instead a change in the ontology: some experiments were taken to show that particles sometimes behave like waves. But particles and waves are incompatible ontologies![10] This *wave–particle duality* seemed crazy, and people tried to get around it. Louis de Broglie introduced a particular wave—the wave function—to account for the behavior of particles. He proposed to associate such wave to each particle as a "guide field" (de Broglie 1928), and Erwin Schrödinger later described the evolution of the wave function by his famous equation. De Broglie's idea was abandoned (perhaps too quickly) on the basis of some criticism by Wolfgang Pauli at the 1928 Solvay Congress. In addition, some other results (such as Heisenberg's uncertainty principle and von Neumann's theorem [von Neumann 1932]) were taken to show that quantum theories *had* to be about the wave function, not about particles. A further problem, however, was that the attempt to interpret quantum mechanics in a realistic fashion as a theory about the wave function seemed to fail. In fact, when Schrödinger tried to do so, he discovered the so-called *measurement problem* (Schrödinger 1935): *if* the wave function completely describes physical systems, *and* it evolves according to the Schrödinger equation, *then* impossible macroscopic superpositions that we clearly never observe (such as the superposition of a living and a dead cat) are produced. Some proposed to solve this problem by introducing the observer actively into the theory: *conscious* observations "collapse" the wave function to one of the terms of the superposition. There are many reasons to consider this approach unsatisfactory, first because of the unfortunate reference to the

---

[10] Particles have definite positions in time, and their temporal evolution is represented by a trajectory in space-time; in contrast, waves are delocalized, spread-out objects that can diffract and interfere with one another.

observer in the formalization of the theory.[11] Be that as it may, the result was that for nearly 20 years everyone gave up on any realistic interpretation of quantum mechanics.

Eventually, in the 1950s new and less problematic proposals to solve the measurement problems were made. Einstein did not like the status of quantum mechanics and proposed an argument to show that the formulation of quantum theory was incomplete and should be supplemented by "hidden variables" (Einstein et al. 1935). Einstein's attempt was unsuccessful, but David Bohm (1952), perhaps with a similar idea in mind, revised and updated de Broglie's particle–wave theory and showed that his theory solves the measurement problem. In Bohm's theory, the description of any physical system is provided by the wave function supplemented by other variables, the particles' positions. In this way, the symmetry among the various terms of the superpositions (dead and living cat) is broken by the presence of the particle trajectories, and the measurement problem is resolved—the cat is dead if the trajectories of the particles composing the cat fall in the support of the dead-cat wave function; she is alive if they fall in the support of the living-cat wave function. However, this theory had an unfortunate fate, since von Neumann's theorem was already taken to prove that hidden variables are impossible. This conviction was reinforced by certain presentations of Bell's inequality, developed in Bell (1964). As a result, Bohm's theory was dismissed for a long time; people believed there was something wrong with it, even if it was not clear what. Only fairly recently was it appreciated that the interpretations of these results were mistaken: it is possible for the quantum world to be made of particles, and there is nothing wrong with Bohm's theory.[12] Still, only a few scholars took the theory seriously, and some of them developed a better formulation that now goes under the name of Bohmian mechanics.[13] Even if there are particles in Bohmian mechanics, people still insisted on the wave function. In fact, the other solutions to the measurement problem focused on either accepting the macroscopic superpositions or eliminating them. Hugh Everett (1957) developed the so called many-worlds interpretation of quantum mechanics, in which the terms of the superpositions are interpreted as belonging to different worlds to which we have no access, so that everything that can happen (all superpositions) will happen, but in a different world.[14] Another possible response to the measurement problem is the GRW theory, proposed by Ghirardi, Rimini and Weber (1986). In the GRW theory, the wave function randomly collapses in one of the terms of the superpositions

---

[11] See Bell 1987, Maudlin 1995, and Goldstein 1998, among others.

[12] For a correct presentation of Bell's theorem, see directly Bell 1964 or Dürret al. 2004, where the so-called no-go theorems against hidden variables theories are also discussed.

[13] See, for example, Dürr et al. 1992, Allori and Zanghì 2004, and Goldstein 2001b for a review of Bohmian mechanics.

[14] For more on the many-worlds theory, see (among others) Vaidman 2002, Wallace 2002, and Barrett 1998.

not because of an observer but as a result of a physical law: the wave function evolves according to a stochastic equation that allows for random spontaneous collapses.[15]

These three examples show how it is possible to provide realist interpretations of the quantum formalism that do not rely on the notion of the observer. For this reason, they have been called *quantum theories without observers* (Popper 1967; Goldstein 1998). Arguing along the lines of the ideas presented in section 1, all these theories were naturally taken to be theories about the wave function, including Bohmian mechanics, which was considered a theory about both the wave function *and* the particles. However, the concern with these theories is that because the wave function lives on configuration space and not three-dimensional space, the explanatory scheme developed in classical theories in terms of a primitive ontology must be drastically revised. A new explanatory scheme is needed, and nobody has found one yet. Hence, Bohmian mechanics, GRW, and manyworlds, as theories of the wave function, at present are not satisfactory theories (see Allori forthcoming).

We can avoid this problem by developing quantum theories with a primitive ontology. Various proposals have been made: they are quantum theories in which, as in classical theories, there is *stuff* in space-time, and we can develop a clear explanatory scheme, along the lines of the classical one, to account for the macroscopic world. As a consequence, there is no quantum revolution (or at least, not the one advertised so far): the quantum world is less crazy and paradoxical than one would have thought. This could be a disappointment for some, but certainly it is a great relief for others—we can still understand things the way we did before! To see where these proposals come from, let us go back to Bohmian mechanics. As we saw, one could think of it as a theory about both particles and the wave function, but if we look closely at its *structure* we see that this approach is contrived. In fact, Bohmian mechanics is naturally a theory with a primitive ontology: there are particles (the primitive ontology), whose temporal evolution is governed by a Schrödinger evolving wave function (the nonprimitive variable). Having understood the role of the wave function in Bohmian mechanics, one can start to look differently to the other quantum theories without observers. The GRW theory as we described it, in which the Schrödinger evolution of the wave function is interrupted by random collapses, does not have a primitive ontology. But two distinct GRW-type theories with primitive ontology have been proposed, originally by Benatti et al. (1995) and Bell (1987), respectively: GRWm, a theory in which the primitive ontology is a field in three-dimensional space defined in terms of the wave function, representing the matter density of physical systems, and GRWf, a theory in which the primitive ontology is a set of discrete points in space-time called "flashes," whose rate depends on the wave

---

[15] For a review of the GRW theory, also called "spontaneous collapse theory," see for instance, Bassi and Ghirardi 2003 and Ghirardi 2002.

function.[16] In this case the primitive ontology is already in space-time, so the set of flashes already provides the set of histories of the primitive ontology. In both GRWm and GRWf, the evolution of the primitive variables is determined by the wave function, which in turns evolves according to the modified GRW dynamics. In addition, Allori et al. (2011) have proposed and developed a many-worlds theory with primitive ontology that they called Sm: a matter density field ontology in three-dimensional space as in GRWm, combined with a Schrödinger evolving wave function that determines the temporal evolution of the primitive variables. A nonexhaustive list of other possible quantum primitive ontologies and their evolutions can be found in Allori et al. (2008).

Because in this framework quantum theories have the same structure as classical theories, in these theories we should be able to recover, at least in principle, all the macroscopic properties of physical objects using an explanatory scheme derived along the lines of the classical one. Indeed, this has been done for Bohmian mechanics in Allori et al. (2002) and Dürr et al. (2004). In the GRW and many-worlds frameworks, more work needs to be done. In any case, see Bassi and Ghirardi (2003) and Goldstein et al. (2011) for some related comments on the matter.[17]

8 Wave Function and Primitive Ontology

What about the wave function? In the primitive ontology framework, the wave function does not represent physical bodies. So what does it do? The role the wave function plays in the theory suggests how we should interpret it. In classical theories, we needed other mathematical entities to implement the evolution for the particles, and here we need the wave function to implement the motion of the primitive variables. This is apparent in Bohmian mechanics, in which the wave function defines the evolution equation for the particles. In GRWm, GRWf, and Sm the situation is analogous: the histories of the primitive variables are determined by the wave function. In Sm the wave function evolves according to Schrödinger equation as in Bohmian mechanics; in GRWf and GRWm it evolves stochastically. In addition, contrary to Bohmian mechanics, in GRWf, GRWm, and Sm, the wave function defines the primitive ontology. That is, in GRWm and in Sm the matter density is given by a certain function $m = m(\psi)$, and, analogously, in GRWf the set of flashes is determined by the wave function.

---

[16] For more on these theories, see, for example, Tumulka 2006 and Allori et al. 2008.

[17] Again, an antireductionist would object to this, but the point here is that in quantum theories with a primitive ontology, we are not worse off than in classical mechanics. That is, whatever objection can be raised against reductionism in classical mechanics could also be raised here in principle; but there is no additional problem for reductionism just due to the fact that we are in the quantum framework.

One could say that because in GRWm, Sm, and GRWf the matter density and the flashes supervene on the wave function, the wave function is "all that is needed," making the primitive ontology superfluous. But that would be a mistake: given a wave function ψ, different ways of defining the matter density and flashes are possible, even if only one is representing physical objects. To determine which one it is, we need more than just the wave function. Consider, for example, among the infinite possible ones, the following simple functions: $m_1 = \psi^2$ and $m_2 = \psi^3$. Both m_1 and m_2 are completely determined by ψ, but ψ alone does not tell us which of the two really represents matter. This additional piece of information needs to be specified in addition to ψ. In other words, the primitive ontology naturally, not logically, supervenes on the wave function, given that there is a law specifying the connection between the primitive ontology and the wave function. Because of this, the information provided by the wave function alone is not enough.

So the role of the wave functions in all these theories is to determine the law of motion for the primitive ontology. In this sense it has a law-like, nomological character. For this reason, Dürr et al. (1992) have proposed that the wave function should be intended as a physical law.

Objections have been raised to this interpretation of the wave function, most vividly by Brown and Wallace (2005). First, laws of nature are time-independent, whereas the wave function, in all quantum theories, evolves itself in time. Dürr et al. (1992) and Goldstein and Teufel (2001) have anticipated and replied to this objection claiming that even if it might be difficult to accept the wave function as a law in the current theories, it will become straightforward once we reach a theory of quantum cosmology in which the wave function is static.

Another objection focuses on the fact that there seem to be multiple degrees of reality: there are material entities, the primitive ontology, and there are nomological entities, represented by the wave function. One could avoid the problem becoming a nominalist with respect to laws. As an alternative, one could maintain that laws exist as abstract entities. One could insist in fact that even if the view has problems, they are not strong enough to make one abandon the view altogether (see Maudlin [2007b] for a recent realist proposal about laws of nature). Another possible option is to try to eliminate the wave function completely from the theory, as has been attempted by Dowker and Henson (2004) and Dowker and Herbauts (2004, 2005).[18]

Note that in classical electrodynamics the electromagnetic fields evolve in time according to Maxwell's equations. Thus, the situation of CEDpf (in which there are fields in the world, in addition to particles) seems very similar in this respect to quantum theories about the wave function: both the wave function

---

[18] Working with a particular GRW model on a lattice, they conjecture that the wave function can be eliminated as a necessary part of the theory. But in other places in their paper they seem to argue that it is not necessary to *know* the wave function to get the correct experimental predictions.

and the fields represent matter, and both evolve in time. The difference is that the electromagnetic fields live on three-dimensional space, not configuration space. Hence, we could consider the fields as describing matter (together with particles) without departing too much from the manifest image, contrary to the situation in quantum mechanics. This is the reason a quantum theory about the wave function is less attractive than a classical electrodynamics about the electromagnetic fields.

9   Symmetry Properties

Before concluding I wish to add a quick remark about the importance of the primitive ontology in connection with the symmetry properties of a theory. Roughly put, a theory is said to be invariant under a given symmetry if the histories of the primitive ontology given by the theory, when transformed under the symmetry, will again be possible histories for the theory.[19] That is, if the original and the transformed histories are both possible solutions of the equations of motion for the primitive ontology, the theory is invariant under that symmetry. The histories of the primitive ontology provide the metaphysical picture of the world, so if the theory is invariant under a given symmetry, this picture *should not change* under the symmetry transformation connected to the symmetry. Given their role, the nonprimitive variables will transform under the symmetry in such a way as to ensure that the histories of the primitive variables are invariant. In other words, because the histories of the primitive ontology need to remain invariant under the symmetry, and given that the evolution of the primitive ontology is determined by the wave function, the wave function will transform in a particular way to make this invariance happen.

Invariance is therefore a property of the dynamics of the primitive ontology: changing the primitive ontology of a theory might change its symmetry properties. So before asking whether a given theory has a given symmetry, it is necessary to identify its primitive ontology and see whether the transformed histories of the primitive ontology are still possible histories for the theory.

Particularly important for quantum mechanics is the question of relativistic invariance: it is usual to assume that a theory is relativistic-invariant if the law of evolution of the wave function is of a particular sort (whether it is a Klein-Gordon or a Dirac equation, for example). But that is a mistake, because the evolution of the wave function is not the thing to look at—whatever the evolution of the wave function is, what is important is the evolution of the primitive ontology. It is worthwhile to mention that the recognition of the importance of the primitive ontology has led to the construction of a relativistic invariant version of GRWf (Tumulka 2006), whereas GRWm still has no relativistic invariant formulation. Relativistic

---

[19] To be more precise, one should mention probability distributions as well. In this regard, see Allori et al. (2008).

invariant single-particle extensions of Bohmian mechanics constructed more or less explicitly with a primitive ontology in mind have been proposed: in Bohm and Hiley (1993) and Dürr et al. (1999) using a wave function evolving according to Dirac's equation, whereas in Berndl et al. (1996) and Nikolic (2005) the wave function used to implement the dynamic of the primitive ontology evolves according to the Klein-Gordon equation.[20] Also, Allori et al. (2011) have developed relativistically invariant extensions of Sm.

10 Conclusion

I conclude by summarizing the common structure of fundamental physical theories based on the notion of primitive ontology:

- Any fundamental physical theory contains a metaphysical hypothesis about what constitutes physical objects: the primitive ontology, which lives in three-dimensional space or space-time and constitutes the building blocks of everything else.
- In the formalism of the theory, the variables representing the primitive ontology are called the primitive variables; in addition, there are other variables necessary to implement the dynamics for the primitive ontology: these variables could be interpreted as (part of) laws of nature.
- Once this is set, one can construct an explanatory scheme based on the one used in classical theories that allows one to determine, at least in principle, all the macroscopic properties of familiar physical objects in terms of the primitive ontology.

This structure holds for classical as well as for quantum theories. Thus, the power of the primitive ontology approach in quantum mechanics is the power of *tradition of clear understanding*, so to speak, given that in this framework many successful ingredients used in classical theories are preserved, such as the essence of its explanatory scheme. In this way, the quantum world ceases to be a mystery, and we can start doing metaphysics through physics as we did so far.

References

Albert, David Z, *Quantum Mechanics and Experience*. Cambridge, Mass.: Harvard University Press (1992).

Albert, David Z, Elementary Quantum Metaphysics. In: J. Cushing, A. Fine, and S. Goldstein (eds.), *Bohmian Mechanics and Quantum Theory: An Appraisal. Boston Studies in the Philosophy of Science* 184:277–284 (1996).

---

[20] These latter attempts have the peculiarity that they seem to involve a particle traveling back in time.

Albert, David Z, *Time and Chance.* Cambridge, Mass.: Harvard University Press (2000).

Albert, David Z, and Barry Loewer, Tails of the Schrödinger's Cat. In: R. Clifton (ed.), *Perspective on Quantum Realities: Non-relativistic, Relativistic, and Field Theoretic* Dordrecht, Kluwer: 81–91 (1996).

Allori, Valia,The Metaphysics of Quantum Mechanics. In: S. Le Bihan (ed.), *La philosopie de la physique: d'aujourd'hui ádemain.* Paris: Edition Vuibert (forthcoming).

Allori, Valia, Detlef Dürr, Sheldon Goldstein, and Nino Zanghì, Seven Steps toward the Classical World. *Journal of Optics* B 4:482–488 (2002).

Allori, Valia, Sheldon Goldstein, Roderich Tumulka, and Nino Zanghì, On the Common Structure of Bohmian Mechanics and the Ghirardi-Rimini-Weber Theory. *British Journal for the Philosophy of Science* 59:353–389 (2008).

Allori, Valia, Sheldon Goldstein, Roderich Tumulka, and Nino Zanghì, Many Worlds and Schrödinger's First Quantum Theory. *British Journal for the Philosophy of Science* 62(1):1–27 (2011).

Allori, Valia, and Nino Zanghì, What Is Bohmian Mechanics? *International Journal of Theoretical Physics* 43:1743–1755 (2004).

Barrett, Jeff, Everett's Relative-State Formulation of Quantum Mechanics. In: *Stanford Encyclopedia of Philosophy* (1998), available online at http://plato.stanford.edu.

Bassi, Angelo, and GianCarlo Ghirardi, Dynamical Reduction Models. *Physics Report* 379: 257–426 (2003).

Bell, John Stuart, On the Einstein Podolsky Rosen Paradox. *Physics* 1(3):195–200 (1964). Reprinted as chapter 2 of *Speakable and Unspeakable in Quantum Mechanics,* Cambridge: Cambridge University Press (1987).

Bell, John Stuart, *Speakable and Unspeakable in Quantum Mechanics.* Cambridge: Cambridge University Press (1987).

Benatti, Fabio, GianCarlo Ghirardi, and Renata Grassi, Describing the Macroscopic World: Closing the Circle within the Dynamical Reduction Program. *Foundations of Physics* 25:5–38 (1995).

Berndl, Karin, Detlef Dürr, Sheldon Goldstein, and Nino Zanghì, Nonlocality, Lorentz Invariance, and Bohmian Quantum Theory. *Physical Review A* 53:2062–2073 (1996).

Bohm, David, A Suggested Interpretation of the Quantum Theory in Terms of "Hidden" Variables, I and II. *Physical Review* 85:166–193 (1952).

Bohm, David, and Basil J. Hiley, *The Undivided Universe: An Ontological Interpretation of Quantum Theory.* London: Routledge and Kegan Paul (1993).

Brown, Harvey, and David Wallace, Solving the Measurement Problem: De Broglie-Bohm Loses Out to Everett. *Foundations of Physics* 35:517–540 (2005).

de Broglie, Louis: La nouvelle dynamique des quanta. In: Solvay Congress 1927, *Electrons et Photons: Rapports et Discussions du Cinquième Conseil de Physique tenu à Bruxelles du 24 au 29 Octobre 1927 sous les Auspices de l'Institut International de Physique Solvay.* Paris: Gauthier-Villars (1928).

Dowker, Fay, and Joe Henson, Spontaneous Collapse Models on a Lattice. *Journal of Statistical Physics* 115:1327–1339 (2004).

Dowker, Fay, and Isabelle Herbauts, Simulating Causal Wave-Function Collapse Models. *Classical and Quantum Gravity* 21:2963–2979 (2004).

Dowker, Fay, and Isabelle Herbauts, The Status of the Wave Function in Dynamical Collapse Model. *Foundations of Physics Letters*18:499–518 (2005).

Dürr, Detlef, Sheldon Goldstein, and Nino Zanghì, Quantum Equilibrium and the Origin of Absolute Uncertainty. *Journal of Statistical Physics* 67:843–907(1992).

Dürr, Detlef, Sheldon Goldstein, Karin Münch-Berndl, and Nino Zanghì, Hyper-surface Bohm-Dirac Models. *Physical Review A*60:2729–2736 (1999).

Dürr, Detlef, Sheldon Goldstein, and Nino Zanghì,Quantum Equilibrium and the Role of Operators as Observables in Quantum Theory. *Journal of Statistical Physics* 116:959–1055 (2004).

Einstein, Albert, Physics and Reality. Reprinted in: A. Einstein, *Out of My Later Years.* New York: Philosophical Library (1950).

Einstein, Albert, Boris Podolsky, and Nathan Rosen, Can Quantum-Mechanical Description of Physical Reality Be Considered Complete? *Physical Review* 47(10):777–780 (1935).

Everett, Hugh, Relative State Formulation of Quantum Mechanics. *Review of Modern Physics* 29:454–462 (1957).

Ghirardi, GianCarlo, Collapse Theories. In: *Stanford Encyclopedia of Philosophy* (2002), available online at http://plato.stanford.edu.

Ghirardi, GianCarlo, Alberto Rimini, and Tulio Weber, Unified Dynamics for Microscopic and Macroscopic Systems. *Physical Review D* 34:470–491 (1986).

Goldstein, Sheldon, Quantum Theories without Observers. *Physics Today* 51(3):42–46 and (4):38–42 (1998).

Goldstein, Sheldon, Boltzmann's Approach to Statistical Mechanics. In: J. Bricmont, D. Dürr, M. C. Galavotti, G. C. Ghirardi, F. Petruccione, and N. Zanghì (eds.), *Chance in Physics: Foundations and Perspectives*: 39–54 (2001a).

Goldstein, Sheldon, Bohmian Mechanics. In: *Stanford Encyclopedia of Philosophy* (2001b), available online at http://plato.stanford.edu.

Goldstein, Sheldon, and Stefan Teufel, Quantum Spacetime without Observers: Ontological Clarity and the Conceptual Foundations of Quantum Gravity. In: C. Callender and N. Huggett (eds.) *Physics Meets Philosophy at the Planck* Scale: 275–289 (2001).

Goldstein, S., R. Tumulka, and N. Zanghì, The Quantum Formalism and the GRW Formalism (2011), available online at http://arxiv.org/pdf/0710.0885.pdf.

Lewis, Peter, Life in Configuration Space. *British Journal for the Philosophy of Science* 55:713–729 (2004).

Lewis, Peter, Interpreting Spontaneous Collapse Theories. *Studies in History and Philosophy of Modern Physics* 36:165–180 (2005).

Lewis, Peter, GRW: A Case Study in Quantum Ontology. *Philosophy Compass* 1:224–244 (2006).

Maudlin, Tim, Three Measurement Problems. *Topoi* 14:7–15 (1995).

Maudlin, Tim, Completeness, Supervenience and Ontology. *Journal of Physics A: Mathematical and Theoretical* 40:3151–3171 (2007a).

Maudlin, Tim, *The Metaphysics within Physics*. Oxford: Oxford University Press (2007b).

Monton, Bradley, Wave Function Ontology. *Synthese* 130:265–277 (2002).

Monton, Bradley, Quantum Mechanics and 3N-Dimensional Space. *Philosophy of Science* 73:778–789 (2006).

Nikolic, Hrvoje, Relativistic Quantum Mechanics and the Bohmian Interpretation. *Foundations of Physics Letters* 18:549–561 (2005).

Popper, Karl, Quantum Theory without "the Observer." In: M. Bunge (ed.), *Quantum Theory and Reality*, Berlin: Springer (1967).

Schrödinger, Erwin, Die gegenwärtige Situation in der Quantenmechanik. *Naturwissenschaften* 23:807–812; 823–823, 844–849 (1935). English translation: John D. Trimmer, *Proceedings of the American Philosophical Society* 124:323–338 (1980).

Sellars, Wilfrid, Philosophy and the Scientific Image of Man. In: R. Colodny (ed.), *Frontiers of Science and Philosophy*, Pittsburgh: Pittsburgh University Press (1962). Reprinted in: *Science, Perception and Reality*, London: Routledge (1963).

Tumulka, Roderich, A Relativistic Version of the Ghirardi-Rimini-Weber. *Journal of Statistical Physics* 125:821–840 (2006).

Vaidman, Lev, Many-Worlds Interpretation of Quantum Mechanics. In: *Stanford Encyclopedia of Philosophy* (2002), available online at http://plato.stanford.edu.

von Neumann, John, *Mathematische Grundlagen der Quantenmechanik,* Springer (1932). English translation: *Mathematical Foundations of Quantum Mechanics*, Princeton, N.J.: Princeton University Press (1955).

Wallace, David, Worlds in the Everett Interpretation. *Studies in the History and Philosophy of Modern Physics* 33:637–661 (2002).

Wallace, David, Everett and Structure. *Studies in the History and Philosophy of Modern Physics* 34:86–105 (2003).

3

Whither Wave Function Realism?*

STEVEN FRENCH

Which features of a theory should a scientific realist take to represent the world? Answer: those that are responsible for the theory's explanatory success. When the theory is quantum mechanics, the wave function is surely one of those features. Indeed, many would regard it as *the* central device of the theory:

> The wave function Ψ of a quantum system completely defines its dynamical state; otherwise stated, all the predictions which can be made concerning the dynamical properties of the system at a given instant of time t can be deduced from a knowledge of Ψ at that instant. (Messiah 1999, p. 61)

Thus, it is from the wave function that we obtain the novel predictions that the realist sets such store by, as well as explanations of a range of quantum phenomena from electron diffraction to quantum tunneling.

Of course, Ψ features in the theory in various ways, where the identification of such "ways" depends on the delineation of the theory and what is included in it. Certainly, Schrödinger's equation, in which Ψ appears perhaps most prominently, would be taken by many as lying at the "core" of quantum mechanics. However, the issues of what else is included in this core and, more generally, what the theory encompasses are delicate and contentious. Some would insist—including some of the architects of the quantum revolution themselves, such as Born, Heisenberg, and Schrödinger, as well as its first philosophical commentators, such as Cassirer—that some of the most significant ontological implications of the theory only emerge when we consider systems of more than one particle (see French and Krause 2006, chapter 3). Constructing an appropriate wavefunction that describes such systems then involves the operation of particle

* This work was supported by a Leverhulme Major Research Scholarship. I thank Alyssa Ney and David Albert for inviting me to contribute to this volume and also the members of the Leeds philosophy of physics group—Maria Kon, Ioan Muntean, Mark Pexton, and Juha Saatsi—for allowing me to express my trenchant views on these and other issues. I particularly thank Kerry McKenzie for helpful comments.

permutations, because the particles are indistinguishable in a nonclassical sense, and thus their "place" in the wave function for the system should not matter. The appropriate mathematical framework here is group theory, introduced in the quantum context in the 1920s and rediscovered in the 1950s. In this respect, what we take to be the theory for realist purposes must extend beyond the one-particle version of Schrödinger's equation to cover what is known as "quantum statistics" and its associated mathematical apparatus (incorporating an appropriately symmetrized form of the relevant many-particle wave function).[1]

This in turn bears on the further issue of how to distinguish what is taken to have physical import from what is regarded as just a mathematical tool or artifact. Thus standardly, as the quote from Messiah makes clear, Ψ is taken to represent the (dynamical) state of the system, and the latter is what the realist might take as describing "how the world is," with Ψ itself understood as merely a convenient mathematical way of capturing it. The wave function realist insists, however, that the wave function be regarded as more than a mathematical artifact. Imbuing Ψ with ontological significance then carries with it a commitment to the (high-dimensional) space the wave function is said to "live" on.

1 Wave Function Realism and Its Motivations

Albert writes,

> The space *we* live in, the space in which any realistic understanding of quantum mechanics is necessarily going to depict the history of the world as *playing itself out* ... is *configuration*-space. And whatever impression we have to the contrary (whatever impression we have, say, of living in a three-dimensional space, or in a four dimensional space-time) is somehow flatly illusory. (1996, p. 277)

This last claim has generated considerable discussion, but there is also the issue of what kind of entity we should take Ψ to be. Again, Albert is clear (following Bell):

> The sorts of physical objects that wave functions *are*... are (plainly) *fields*—which is to say that they are the sorts of objects whose states one specifies by specifying the values of some set of numbers at every point in the space where they live. (1996, p. 278)

---

[1] This is constructed from the nonsymmetrized form via the action of the permutation group. The two most well-known constructions are the symmetric, corresponding to Bose-Einstein statistics, and the antisymmetric, corresponding to Fermi-Dirac statistics, although other forms, corresponding to parastatistics, are also theoretically possible.

Before considering these issues—how we recover our impressions and how we should understand fields—let us first consider three motivations for adopting wave function realism. I shall suggest that it is the third motivation that carries the most force but that it encourages us to broaden our ontological scope.

The first motivation insists that regarding the wave function as real is essential for solving the infamous measurement problem, on whatever interpretation of the theory we choose (Bohmian, Everettian, Ghirardi-Rimini-Weber [GRW], and so on). However, even granted the crucial role played by Ψ—by, for example, describing the set of appropriate worlds in the Everett interpretation—this is not a particularly strong motivation for taking Ψ realistically. One could argue that it and the accompanying configuration space are mere mathematical devices for conveniently representing the set of worlds, say, and this does the actual work in the resolution. Indeed, with an eye on the issue of how we recover our impressions, the Everettian might insist that the configuration space framework gets in the way of understanding what is going on, because according to her account, "I" am indexed to a particular world and as such, I do not "live" in configuration space in the relevant sense. What that space seems to represent on this interpretation are all the I's, which as a collective might be said to live in it, but individually, each lives in its particular world or branch, describable in familiar three-dimensional terms.

This brings us to the second motivation, which has to do with entanglement.[2] Thus it is argued that entangled states cannot be adequately characterized in terms of states of entities living in familiar, three-dimensional space; rather, they must be characterized as states of something—namely, the wave function—spread out across a higher-dimensional (configuration) space (see Ney 2012, p. 533). However, the foregoing claim is disputable. An alternative metaphysics for adequately characterizing such states can be elaborated in terms of the notion of "nonsupervenient" relations holding between individual objects existing in familiar three-dimentional space (Teller 1986; French and Krause 2006). Comparing the costs of alternative understandings of entanglement is tricky, and the issue of what counts as an "adequate characterization" will loom large in such a comparison, but the very existence of such alternatives reduces the force of this motivation.

The third, broader motivation is that wave function realism offers the most transparent reading of those versions of quantum mechanics that offer a solution to the measurement problem:[3]

---

[2] Schrödinger famously took *this* to be the defining feature of quantum mechanics (1935, p. 555).

[3] I leave to one side concerns that such solutions might be viewed not as part of the theory but as arising from interpretations of it. To respond to them would obviously require explicating the distinction between "theory" and "interpretation."

> On the most straightforward, ontological understanding of all of [the] realist versions of quantum mechanics, we have at least one law, the Schrödinger equation, that describes the behavior of at least one unfamiliar entity: the wavefunction. (Ney 2012, p. 532)[4]

Here there might be a concern about the relationship between the laws and the ontology that is read off the theory. In the case of Newton's Second Law, by contrast (Ney 2012), we do not take the acceleration as an element of our fundamental ontology, but as representing the change of state of the object with the given mass, understood as a state-independent property. In that of Schrödinger's equation, it might appear that the principal unfamiliar entity here is Ψ, thus encouraging a straightforward realist reading. However, here too we have state-independent properties, such as (rest) mass, spin, and charge, only they are tucked away off-stage (in the wings, as it were). Bringing them into the spotlight allows us to appreciate their role in pinning down the kind of particle, say, whose state is encoded in Ψ. Recognizing this role will cause us to enlarge our ontological scope and read off our ontology from a broader set of features of the theory, beyond Schrödinger's equation. In what follows I will briefly explore this expanded reading off.

Before I do so, let us consider the ontological picture arrived at by taking the above motivation seriously. It obviously presents two features: the wave function itself and the space on which it "lives." Albert is clear, as we have seen, that the wave function should be understood as a field. Now, as Redhead (1995) pointed out, there exists a kind of underdetermination of the metaphysics by the physics here, in the sense that the physics supports the following two alternatives (see French and Krause 2006, pp. 51–54): the fields in question can be regarded either as global particulars, ontologically independent of the points of the space, but with some correspondence existing between the points and the field quantities; or one can take them, or rather the field quantities, as nothing but properties of the points of the space.

Taking the former option would ultimately yield only one global particular— the "universal" wave function.[5] Here one might turn to the monist's wardrobe for a possible metaphysical suit of clothes with which to dress this view. One such would be Horgan and Potrc's (2008) "blobjectivism": leaving to one side the details of their "austere realism," their "blob" has all the characteristics of a field-like global particular, where the underlying thought seems to be that a field can be regarded as a vast universe-spanning substance, or "jello-stuff." Indeed, Horgan and Potrc suggest that not only is it a "conceptually coherent possibility

---

[4] Note, however, that there is some tension here with the first, narrower motivation, because the Schrödinger equation cannot accommodate measurement, of course.

[5] Strictly speaking this would not be a monistic view, given the role played by the underlying space.

that the actual world we humans inhabit does not have any real parts" (2008, p. 171) and thus can be metaphysically understood via blobjectivism, but that "this is an attractive-looking ontological framework for physics, especially if one focuses on broadly field-theoretic formulations of physical theory" (2008, p. 171). Of course, whereas they construe the "entire cosmos" in blobjectivist terms as a physical field, along the lines presented by quantum field theory,[6] say, and thus as living in "standard" 3 + 1 space-time, the wave function realist would take the quantum mechanical blob as sitting on a deeper level, as it were, in configuration space.

Alternatively, a field can be understood in terms of properties of the points of the space, so the latter is the fundamental entity and it stands in the same relation to the field as a subject does to its predicates. Adopting this option, the wave function might be seen not as some global blob but in terms of an assignment of properties to the points of configuration space. In this case, again, some account of the metaphysics of that space would need to be given. The worry here might be that one is then driven into the arms of some form of configuration space substantivalism. Now substantivalism is famously problematic in the space-time context, but it is not clear that the same problems would arise here. Nevertheless, to insist that the ontology of the world is a substantival configuration space bearing the relevant properties makes apparent the "radically revisionary nature" of wave function realism that so distresses certain commentators (Monton 2006). I shall not go into this further here because my interest is to simply highlight the underdetermination that arises in this context. I shall return to the implications of this underdetermination for wave function realism later, but on either of the foregoing options, some consideration has to be given as to how we recover the apparently "illusory" appearances.

Hawthorne (2010) identifies two approaches to this recovery: the "conservative," which identifies features of the appearances, such as "everyday" objects, with aspects of the fundamental ontology; and the "liberal," which posits generational principles by means of which one obtains the appearances from the fundamental ontology. Taking the example of classical field theory and our two underdetermined alternatives, we obtain the following options. With the field understood as a "blob," or global particular, the conservative would identify everyday objects with certain features or aspects of this blob, whereas the liberal would take these features to "lead to" the existence of these objects, where that leading to can be articulated in terms of some generational principle or other. Alternatively, where the field is understood in terms of an assignment of values at points in (Galilean) space-time, the conservative would identify everyday objects with regions of space-time, whereas the liberal would take the existence

---

[6] In a recent presentation at the Parts and Wholes Workshop in Leiden, Richard Healey also offered blobjectivism as a metaphysical interpretation of quantum field theory (Healey forthcoming).

of certain patterns of field quantities over space-time to lead to the existence of such objects, with, again, the above articulation (Hawthorne 2010, p. 146).

Now there are concerns that can be raised with regard to both the conservative and liberal approaches but Hawthorne has a meta-worry: assuming that we can recover the appearances, how do we distinguish those bases for such a recovery that are absurd or otherwise intolerable from those that are not? One option is to focus on any explanatory gaps in such recoveries (Hawthorne 2010, pp. 148–51). One can then attempt to dissolve such a gap or live with it. The problem, according to Hawthorne, is that it may not be clear which gaps are "livable" and which are not. Critics of wave function realism have argued that there are clear "unlivable" gaps in the recovery of "everyday" space from configuration space (see, for example, Monton 2006).

An alternative heuristic is to adopt some metasemantical principle or other, such as that associated with the causal theory of reference, or Lewis's suggestion that predicates "semantically gravitate" to the more natural properties in the world (Hawthorne 2010, p. 151). Thus, Hawthorne deploys this heuristic to raise a further concern about the following line of reasoning that the wave function realist may indulge in:

> Assume that all there is to the world is configuration space; then the best package, all things considered, is one that has ordinary macro-predicates pick out features of configuration space. But this shows that certain features of configuration space are *good enough* to count as tigers. (2010, p. 151)

In that case, the reasoning continues, even if one adds "extra stuff" such as material particles, one should still count the relevant features of configuration space as tigers. Hawthorne regards this as "dicey," because one might end up associating the predicate "tiger" with some kind of gerrymandered properties, whereas a "richer metaphysics" might trump that claim by associating tigers with more "natural" properties, and the Lewisian heuristic strategy would push one to adopt the latter metaphysics.

However, it is not clear how one might draw a firm distinction between "gerrymandered" and "natural" properties in this context. Wave function realism offers a fundamental ontology grounded in modern physics, and if the source of what counts as "natural" is drawn from our experience of everyday objects, the Lewisian strategy will obviously conflict with the aims of this program. If, on the other hand, this source is taken to be our fundamental theories of physics, then advocates of wave function realism will insist that the worry about picking out gerrymandered properties evaporates, because the basis of their ontology is "natural" in this sense—as we have seen this is precisely one of the motivations.[7]

---

[7] For a more developed response to Hawthorne, see Ladyman 2010.

When it comes to the concern about explanatory gaps, two obvious questions arise: what is it that we are looking to be explained? And what sort of explanation would be acceptable?

With regard to the first question, if the aim is that of recovering the appearance of definite, spatially localized, solid (in some cases) macro-level objects,[8] then the relevant physics can take us some way toward this. Definiteness is taken care of via decoherence, by which we "recover" definite position measurements, for example (see Hartle 2010). Solidity is accounted for, and thereby recovered, via Pauli's Exclusion Principle, which in turn arises from the antisymmetrization, under particle permutations, of the relevant wave functions for fermionic systems (as touched on above). And so we might continue ...

With regard to the second question, one might insist that the foregoing physics-based recovery not only yields all the explanation we are going to get, it provides all the explanation we need (see for example, Ladyman 2010, p. 159). In this case, the worry about explanatory gaps can be dismissed as based on inappropriate comparisons with other relationships where some form of recovery is required. Certainly, the quantum-classical relationship seems disanalogous to that which holds between the mind and the brain, for example (Hawthorne 2010, p. 150). In the latter case, certain compelling metaphysical understandings of the nature of the mind lead to the appearance of such an explanatory gap, whereas in the former, analogous understandings are either absent or, at least, less compelling. With no such gaps, it seems that the physics does all the explanatory work in these cases. Whether the *only* role for metaphysics is to unify such explanations under some appropriate relationship—such as "grounding" or "dependence"—is a contentious issue, but appealing to the relevant physics may certainly gut the metaphysics of much of its usefulness. If the way certain properties of a table, for example, depend on certain features of the micro-world is cashed out entirely in terms of quantum mechanics, then we can ask: what work is the notion of dependence actually doing here? This concern might be dismissed as a result of taking naturalism too seriously, but however one conceives the role of metaphysics in general, the relevant metaphysical principles need to be articulated in such a way that they can be seen to be applicable to these kinds of cases.

Returning to blobjectivism, it is in this regard that such an account falls short. According to Horgan and Potrc, one cannot say that physical magnitudes, in all their huge variety, are simply instantiated by *parts* of the field, because as a monadic blob, the field has no parts. Instead, they introduce an unrestricted plethora of "manners of instantiation," in the sense that the blob itself, in its entirety, as it were, instantiates in a certain manner (and, in particular, in a

---

[8] As expressed by Allori et al. 2008, p. 370, for example, in the context of a comparison of Bohmian mechanics and the GRW theory, where the role of the wave function is to govern the motion of matter.

spatiotemporally local manner) the relevant properties and relations (Horgan and Potrc 2008, p. 169). However, the lack of clarity that surrounds these "manners of instantiation" is, at the very least, unhelpful.[9]

Alternatively one might appeal to some form of "grounding" relation as the appropriate "umbrella" metaphysics: *a* is said to be grounded in *b* in the sense that *a* holds in virtue of *b*, without it being the case that only *b* exists. Thus, the property of solidity (of our table, say) might be said to be "grounded" in the symmetry features of the relevant wave function and, more generally, three-dimensional space, and "everyday" objects can be said to be grounded in configuration space and the wavefunction. Truths about such everyday objects would then hold *in virtue of* some other, more fundamental, "facts" about the wave function and configuration space (North 2010, p. 26).

The worry here, as indicated, is that the kind of dependence that "in virtue of" signifies effectively evacuates all there is to *a* in favor of the relevant features of *b*. If all there is to *a* is explained in terms of features of *b*, then what is left that has any independent existence?[10]

Answering "nothing" would be to understand the "recovery" of the appearances in "eliminativist" terms. This is certainly how Albert originally saw it, in the passage reproduced above. Such eliminativism seems to make people nervous,[11] perhaps because of the implicit acceptance of a form of error theory according to which our claims about the appearances would be simply false. However, perhaps the concern here might be assuaged by adopting an alternative strategy. Thus, one could still reject tables, people, everyday objects in general as elements of one's fundamental ontology, while continuing to assert truths about them. One could, for example, follow Horgan and Potrc (2008) again and adopt a context-based distinction between truth as direct and indirect correspondence, where the latter is understood as semantic correctness under the appropriate contextually operative semantic standards. Take the statement "There are tables." This is true, in the "indirect correspondence" sense, under those contextually operative standards governing "ordinary" usage. Of course, these are not the standards appropriate for the context of "serious ontological enquiry," where this context will embrace the wave function and configuration space. Here standard, "direct correspondence" would be appropriate.

One might balk at contextualizing truth in this way, so here is another option: retain our standard semantics but allow "simples" of some form or other to act as truth-makers of "There are tables" (Cameron 2008). Then introduce

---

[9] Thus one might prefer a further metaphysical understanding of fields articulated via an appropriate notion of parthood. Of course, one then has to say something about how such parts compose, something Horgan and Potrc precisely wish to avoid with their idea of manners of instantiation.

[10] Here talk of "facts" and their being real but nonfundamental may obscure the issue.

[11] "The pedant in me forces me to note that it's probably better to say that on this view three-space, like Eddington's table, is not *fundamental*, rather than being illusory" (Callender 1998).

a distinction between the language used to describe "how the world is at its fundamental level" (Cameron 2008, p. 300) and the language in which statements such as "There are tables" are expressed. If we distinguish the former by using bold type, then "**there are tables**" is false, because at the fundamental ontological level there are no tables. However, "there are tables" is true, and not in virtue of the fact that **there are tables** but in virtue of the fact that there are simples that are arranged table-wise (Cameron 2008, p. 301). In the present context, we might thus insist that "there are tables" is true, in virtue of the fact that the wave function can be, under appropriate circumstances, "arranged table-wise." What would be the simples in this case? Well, taking the view of the wave function, *qua* field, as substantival or blob-like, it would be the wave function itself that is regarded as a "simple." Now this might seem odd, given that it packs a lot of complexity. Furthermore, if we take a simple to be a (metaphysical) object that has no proper parts, then arguably that complexity indicates that the wave function does have parts (of a kind).[12] If we further insist that simples are not spatially extended, then, again arguably, the wave function is not simple, although of course the space it extends across is configuration space. Alternatively, if we understand it in terms of field quantities instantiated at configuration space points, then we can take these—field quantities-instantiated-at-configuration-space-points—as our simples. Neither the points nor the quantities should be taken as "parts," properly speaking, because the relationship here is not that of parthood but of instantiation, proceeding in one direction, and bearing (as in a substance bearing properties) proceeding in the other.

All of the above metaphysical moves come with costs of one form or another. Ultimately one has to make one's choice and pay the price. What is important is that there is no need, on any of these schemes, to give up most of our typical judgments, beliefs, and so on or true talk about everyday objects and the world of appearances, which seems to be a major stumbling block for many commentators on wave function realism. With that in mind, let us now return to the nature of the ontology proposed.

Unfortunately this remains unclear, given the underdetermination with regard to the notion of "field." The physics itself gives no grounds for choosing between the "blobby" particular and "instantiated quantities" options. One might appeal to some form of metaphysics-based argument to advance one over the other. We recall, for example, that blobjectivism incorporates an unrestricted plethora of spatiotemporal, local "manners" of instantiation, which allow for the accommodation of the complexity and variety of the appearances within the monadic ontology of the "blob." One might then point to the lack of clarity over these "manners of instantiation" as discrediting the blobjectivist understanding of the field-as-global-particular and thereby leaving only the interpretation of the wave function as field quantities-instantiated-at-configuration-space-points.

---

[12] Although, as we have already seen, Horgan and Potrc̆ would resist this indication.

However, there are other metaphysical forms that could be tailored for the former alternative, and other generational principles that could be invoked. More worryingly, perhaps, appealing to further metaphysics in this way leaves the realist open to criticism from the antirealist (see van Fraassen 1989).

Relatedly, such an appeal would involve a significant extension of the degree of epistemic humility that would have to be adopted with regard to our understanding of how the world is. Such humility arises as a result of the existence of an extensive array of metaphysical "facts" about which we can have no knowledge, and the broader that array, the deeper the attitude of humility we must adopt. Thus, consider the following motivational argument: we can have knowledge of something only insofar as it affects us, so our knowledge is dependent on certain relations holding. These relations are not supervenient on or otherwise reducible to the intrinsic properties of things; hence, we must remain ignorant of and adopt an attitude of humility toward these intrinsic properties.[13] Perhaps it is impossible to avoid being humble to some degree, but certainly the more metaphysics we introduce into our realism, the less firmly grounded this understanding would be in our current best science and the more humble an attitude we must adopt (see Langton 2004). However, we can (and should) strive to reduce our level of humility as much as we can (if only to avoid the empiricist's scorn!). Redhead's own response to the underdetermination in the interpretation of field theory was, in effect, to urge such a reduction by articulating a view of fields in structuralist terms (see French and Ladyman 2003). Thus, he took this view as hinging on the difference between the questions "what is a field?" and "what are the equations which govern its behaviour?" (Redhead 1995, p. 18). The first can only be fully answered by appealing to the kinds of metaphysical understandings that blobjectivism and configuration space substantivalism provide, but, as Redhead emphasizes, this answer is not exhausted by the answer to the second question. I suggest that the wave function should be considered likewise. This not only deals with the underdetermination of ontology but also accommodates the further important features of quantum theory noted earlier, such as the role of particle permutations and the associated symmetries.

These features are typically articulated in terms of the Hilbert space representation of quantum mechanics that is rejected by the wave function realist in favor of configuration space. On the one hand, the motivations for this rejection are, again, not particularly compelling; on the other hand, the wave function realist can accommodate the foregoing features, so she can press the point that the costs involved in adopting her stance are reduced. I shall suggest that what is doing the work in both cases is group theory and that the wave function realist would be better off cloaking herself in a form of structuralism.

---

[13] This is a crude condensation of the argument given in Langton 1998.

2 Hilbert Space: Representation and Structure

It was von Neumann's great insight that wave and matrix mechanics could be understood as equivalent through the introduction of Hilbert space (see, for example, Kronz 2004; for a detailed analysis, see Muller 1997). Subsequently, this too has come to be seen as the "heart and soul" of quantum mechanics (Ismael 2009). Essentially what von Neumann did was to specify a set of functions that could instantiate such a space and could be identified with Schrödinger's wave functions. Structural realists have taken this move to be ontologically significant, insisting that the structure of the world is, in some sense, Hilbert space structure (Ladyman 1998). Now, the wave function realist could insist that von Neumann's identification was merely mathematical and not ontological and thus resist the lesson from history, arguing that Hilbert space is simply a mathematical device that provides a convenient representation (see North's chapter in this volume).[14] The Hilbert space framework itself might also be rejected as abstracting away too much structure, thereby leaving us without a perspicuous picture of the world, where perspicuity is understood in terms of our familiar ontology of particles and fields (see North's chapter).

However, this familiarity is a bit of a sham. After all, how can we say what the world is like in these terms if we can't say whether the particles are individuals or not (French and Krause 2006), or the fields are blob-like particulars or properties of space points? Again, we need to be careful with our ontological claims. Furthermore, one might be inclined to retain Hilbert space as one's ontological "arena" because it supports the group-theoretic representations in terms of which we understand, for example, the effect of particle permutations and, consequently, the Exclusion Principle, the distinction between Fermi-Dirac and Bose-Einstein statistics, as well as the nonclassical nature of "spin". Basically, the relevant group—such as the permutation group—induces a decomposition of the Hilbert space into invariant subspaces, in each of which the group has an irreducible representation. This yields considerable information that is independent of the specific dynamics, something that was seized on as the only way of dealing with the computational intractability of anything but the simplest situation. Thus, it is through the representations supported by Hilbert space that group theory captures "the essential features which are not contingent on a special form of the dynamical laws nor on special assumptions concerning the forces involved" (Weyl 1931, p. xxi).

Of course, the configuration space representation can also accommodate quantum statistics and properties such as spin: in this context, the action of the particle permutations identifies those points of the space that correspond to such permutations, yielding the reduced quotient space (for a brief discussion

---

[14] Of course, something further needs to be said as to why this should be taken as artifactual and not Ψ itself.

see French and Krause 2006, pp. 176–78). Within this one can then obtain fermionic and bosonic statistics, as well as more exotic forms (Imbo, Shah Imbo, and Sudarshan 1990).[15] Furthermore, configuration space can also be used to underpin the connection between spin and these particle statistics (see Berry 2002). The important point is that what is doing the work here, and in many applications of quantum theory, is ultimately the kind of structure that group theory captures, and this is what the wave function realist needs to incorporate into her picture. If she does, and takes seriously the challenge represented by the forms of underdetermination mentioned above, she will inevitably move to a more structuralist stance, albeit one whose commitments are cashed out in terms of configuration space rather than Hilbert space (see Ladyman 2010).

This will allow her to reduce the level of metaphysical humility that arises from positing more ontology than our epistemology can support. She can resolve the above underdetermination by dispensing with the object-oriented ontologies that underpin the apparent alternatives—fields as "blobs" (that is, global objects) or as properties of space points (where the points are the objects)—and instead invest the relevant structure with ontological significance (French and Ladyman 2003). In general, this structure is constituted by the laws and associated symmetry principles of our fundamental theories (French 2006). In particular, in this case, it will encompass Schrödinger's equation. From the structuralist perspective, this should not be seen as governing the evolution of the wave function *qua* object;[16] rather, it expresses the dynamic nature of the structure itself. Returning to the issue of what features of our theories the realist should take as ontologically significant, we see that this illustrates a further aspect of the structuralist shift: for many realists, laws are a guide to the fundamental nature of a world, but for the structuralist, laws *are* the fundamental nature of the world (Cei and French forthcoming). Furthermore, the symmetries are encoded in the group-theoretic structure, and this will still allow the wave function realist to construct putative objects via the representation of physical quantities in terms of sets of invariants (French and Ladyman 2011). Thus in the case of hadrons, for example (those particles that are now understood to be composed of quarks and of which the most well-known examples are protons and neutrons), from a knowledge of the relevant structure we not only can determine all the relevant fundamental particles but can also obtain an appropriate composition relation in terms of which the full hadron ontology can be obtained (McKenzie forthcoming).

The metaphor that is often invoked here is of objects as mere "nodes" in the structure, and here we can replay the discussion about generational principles and ask if such reductive moves should be taken to imply that the objects—including,

---

[15] The relevant groups are "braid" groups.

[16] It has been argued that the notion of "governance" here is deeply problematic (Mumford 2004).

here, elementary particles—do not exist. At this point the realist can adopt one or another of the metaphysical maneuvers already indicated and insist that we can retain all our judgments and beliefs about such objects, but that fundamentally what exists is the relevant underlying structure; that is, elementary particles do not exist *qua* particles-as-objects, what exists is the structure represented by symmetry groups such as SU(3) in terms of which the relevant physical quantities are invariant.

Within this framework we can also "recover" ordinary notions such as position (Castellani 1998) and given that, there seems to be no greater loss of perspicuity than in the standard, "object-oriented" wave function realist's picture. Furthermore, as I have indicated, one need not give up the emphasis on the significance of Schrödinger's equation. This broader perspective may also allow the wave function realist to accommodate the role of symmetry principles in physics, where the relevant symmetries are understood as features of the world-as-structure over and above that represented by laws such as Schrödinger's. In particular, what needs to be accommodated is Wigner's fundamental insight that the representations of the Poincaré (or inhomogeneous Lorentz) group characterize the (free) relativistic particles (see McKenzie forthcoming; Mirman 1995, chap. 6). In particular, a classification of these representations yields the possible relativistic wave equations, including the Dirac equation for massive spin 1/2 particles, where these properties are given by the eigenvalues of the Poincaré invariants. From the Dirac equation we can obtain Schrödinger's equation in the nonrelativistic limit, and by going to the classical limit we get Newton's Second Law, so that both the latter are seen as "direct requirements" of the Poincaré group (Mirman 1995, p. 118). Of course, the usual interpretation of the Poincaré group is in terms of the symmetries of Minkowski space-time, so the challenge for the wave function realist now is to articulate how the latter may be "recovered" in such a way that the symmetries are still accorded a fundamental role (see Wallace and Timpson 2010, p. 24).

3 Conclusion

The wave function realist takes only a certain part of quantum mechanics—namely, that represented by Schrödinger's equation—and reads it in a certain way—namely, as governing the evolution of an object, the wave function. This initial move leaves out much that is powerful in quantum mechanics, particularly from the group-theoretic perspective. By treating the wave function as a field, a form of underdetermination arises, the adoption of either horn of which introduces unnecessary metaphysical humility.

Accommodating symmetries and reducing humility push the realist in the structuralist direction. From this perspective, the wave function is neither an object nor a mere mathematical artifact. Mathematically it encodes and thus represents the relevant properties standardly attributed to putative objects but which for the structuralist are aspects of or emerge from the structure of the world. The expression of its evolution via Schrödinger's equation is not the expression of the evolution of a physical object in its own right, but that of the essential dynamical nature of this structure. The structuralist needs to understand the nature of the wave function in order to grasp this dynamical nature, but the wave function does not need to be objectified for us to understand quantum mechanics. Wave function realism gives us a transparent understanding of a core feature of quantum physics, but it does so at a cost that goes beyond that of bringing a multidimensional space into the picture. A structuralist account offers us a less transparent view, perhaps, but it allows us to see further.

References

Albert, D. Z. (1996), "Elementary Quantum Metaphysics" in J. T. Cushing, A. Fine, and S. Goldstein (eds.), *Bohmian Mechanics and Quantum Theory: An Appraisal*. Dordrecht: Kluwer, pp. 277–284.

Allori, V., S. Goldstein, R. Tumulka, and N. Zanghí (2011). "Many Worlds and Schrodinger's First Quantum Theory," *British Journal for the Philosophy of Science* 62:1–27.

Berry, M. (2002). "Indistinguishable Spinning Particles", in A. Fokas, A. Grigoryan, T. Kibble, B. Zegarlinski (eds.), *XIIIth International Congress of Mathematical Physics*. Boston: Boston International Press, pp. 29–30.

Callender, C. (1998), "Review of J.T. Cushing, A. Fine and S. Goldstein (eds), *Bohmian Mechanics and Quantum Theory: An Appraisal*. Dordrecht and London, Kluwer Academic Publishers, 1996," *British Journal for the Philosophy of Science* 49:332–37.

Cameron, R. (2008), "Truthmakers and Ontological Commitment," *Philosophical Studies* 140:1–18.

Castellani, E. (1998), "Galilean Particles: An Example of Constitution of Objects," in E. Castellani (ed.), *Interpreting Bodies: Classical and Quantum Objects in Modern Physics*. Princeton, N.J.: Princeton University Press, pp. 181–94.

Cei, A., and S. French (forthcoming), "Getting Away from Governance: Laws, Symmetries and Objects," available online at http://philsci-archive.pitt.edu/5462.

French, S. (2006), "Structure as a Weapon of the Realist," *Proceedings of the Aristotelian Society* 106:167–85.

French, S., and D. Krause (2006), *Identity and Individuality in Classical and Quantum Physics*. Oxford: Oxford University Press.

French, S., and J. Ladyman (2003), "Remodelling Structural Realism: Quantum Physics and the Metaphysics of Structure," *Synthese* 136:31–56.

French, S., and J. Ladyman (2011), "In Defence of Ontic Structural Realism," in P. Bokulich and A. Bokulich (eds.), *Scientific Structuralism*, Boston Studies in the Philosophy of Science. Dordrecht: Springer, pp. 25–42.

Hartle, J. (2010), "Quasiclassical Realms," in S. Saunders, J. Barrett, A. Kent, and D. Wallace (eds.), *Many Worlds? Everett, Quantum Theory, and Reality*. Oxford: Oxford University Press, pp. 73–98.

Hawthorne, J. (2010), "A Metaphysician Looks at the Everett Interpretation," in S. Saunders, J. Barrett, A. Kent, and D. Wallace (eds.), *Many Worlds? Everett, Quantum Theory, and Reality*. Oxford: Oxford University Press, pp. 144–53.

Healey, R. (forthcoming), "A Lego Universe? The Physical Construction of the World." Lecture presented at the Lorentz Center, Leiden, March 24, 2010.

Horgan, T., and M. Potrc (2008), *Austere Realism: Contextual Semantics Meets Minimal Ontology.* Cambridge, Mass.: MIT Press.

Imbo, T. D., C. Shah Imbo, and E. C. G. Sudarshan (1990), "Identical Particles, Exotic Statistics and Braid Groups," *Physics Letters* B 234:103–7.

Ismael, J. (2009), "Quantum Mechanics," *Stanford Encyclopedia of Philosophy*, available online athttp://plato.stanford.edu/entries/qm/.

Kronz, F. (2004), "Quantum Theory: Von Neumann vs. Dirac," *Stanford Encyclopedia of Philosophy*, available online at http://plato.stanford.edu/entries/qt-nvd/.

Ladyman, J. (1998), "What Is Structural Realism?," *Studies in History and Philosophy of Science* 29:409–24.

Ladyman, J. (2010), "Commentary. Reply to Hawthorne: Physics Before Metaphysics," in S. Saunders, J. Barrett, A. Kent, and D. Wallace (eds.), *Many Worlds? Everett, Quantum Theory, and Reality.* Oxford: Oxford University Press, pp. 154–60.

Langton, R. (1998), *Kantian Humility. Our Ignorance of Things in Themselves.* Oxford: Oxford University Press.

Langton, R. (2004), "Elusive Knowledge of Things in Themselves," *Australasian Journal of Philosophy* 82:129–36.

McKenzie, K. (forthcoming), "Structuralism and Intrinsics."

Messiah, A. (1999), *Quantum Mechanics.* Dover: Wiley and Sons, 1958.

Mirman, R. (1995), *Group Theory: An Intuitive Approach.* Singapore: World Scientific.

Monton, B. (2006), "Quantum Mechanics and 3N-Dimensional Space," *Philosophy of Science* 73:778–89.

Muller, F. (1997), "The Equivalence Myth of Quantum Mechanics," *Studies in the History and Philosophy of Modern Physics* 28:35–61, 219–47.

Mumford, S. (2004), *Laws in Nature.* London: Routledge.

Ney, A. (2012), "The Status of Our Ordinary Three Dimensions in a Quantum Universe." *Noûs* 46:525–560.

Redhead, M. L. G. (1995), *From Physics to Metaphysics.* Cambridge: Cambridge University Press.

Schrödinger, E. (1935) "Discussion of Probability Relations between Separated Systems," *Proceedings of the Cambridge Philosophical Society* 31:555–63.

Teller, P. (1986), "Relational Holism and Quantum Mechanics," *British Journal for the Philosophy of Science* 37:71–81.

van Fraassen, B. (1989), *Quantum Mechanics: An Empiricist Approach.* Oxford: Oxford University Press.

Wallace, D., and C. Timpson (2010), "Quantum Mechanics on Spacetime I: Spacetime State Realism," *British Journal for the Philosophy of Science* 61:697–727.

Weyl, H. (1931), *The Theory of Groups and Quantum Mechanics*, 2nd ed. London: Methuen.

4

Reality and the Role of the Wave Function in Quantum Theory

SHELDON GOLDSTEIN AND NINO ZANGHÌ

1 Questions about the Wave Function

We are concerned here with the role and status of the wave function in quantum theory, especially in Bohmian mechanics. What we describe is joint work with Detlef Dürr.

The wave function is arguably the main innovation of quantum theory. Nonetheless, the issue of its status has not received all that much attention over the years. A very welcome step is this book, whose main concern is the question, What is this strange thing, the wave function, that we have in quantum mechanics? What's going on with that? Who ordered that?

In more detail, is the wave function subjective or epistemic, or is it objective? Does it merely describe our information, or does it describe an observer-independent reality? Why does the wave function collapse? What's going on there? If the wave function is objective, is it some sort of concrete material reality or something else?

Let us say a word about what it means for the wave function to be merely epistemic. To us that means first that there is something else, let's call it X, describing some physical quantity, say, the result of an experiment, or maybe the whole history up to the present of some variable or collection of variables—things we're primarily interested in. Then to say that the wave function is merely epistemic is to say that it is basically equivalent to a probability distribution on the space of possible values for X.

You should note that orthodox quantum theory is not of this form. That's because the X is in effect a hidden variable and there are no hidden variables in orthodox quantum theory—there's just the wave function. So the wave function is certainly not merely epistemic in orthodox quantum theory.

And neither is Bohmian mechanics (Bohm 1952; Bell 1966; Goldstein, 2001) of this form. In Bohmian mechanics, it is indeed the case that the wave function sort of has a probabilistic role to play, because the absolute square of the wave function gives the probability of the configuration of the Bohmian system. However, that's not the only role for the wave function in Bohmian mechanics; it's not its fundamental role and certainly not its most important role.

We should all agree—and maybe this is the only thing we would all agree on—that there are three possibilities for the wave function. (1) It is everything, as would seem to be the case with Everett (1957). (2) The most modest possibility: it is something (but not everything), as with Bohmian mechanics, for example, where there's the wave function and something else. Or (3) maybe it's nothing—which would solve the problem of having this weird thing, the wave function: if you can get rid of it, you don't have to agonize about it.

2 Bohmian Mechanics

Let's turn to Bohmian mechanics, for us the simplest version of quantum mechanics. In Bohmian mechanics you have for an N-particle system the usual quantum mechanical wave function $\psi = \psi\,(\mathbf{q}_1, \ldots, \mathbf{q}_N)$—in the simplest case, of spin-0 particles, a complex-valued function of the "generic positions" of the particles—but it's not everything: in addition to the wave function you have the actual positions of the particles, $\mathbf{Q}_1, \ldots, \mathbf{Q}_N$, which form the configuration Q.[1]

We say that the positions of the particles provide the primitive ontology of the theory (Dürr, Goldstein, and Zanghì 1992; Goldstein 1998; Allori et al. 2008). In so saying, we wish to convey that the whole point of the theory—and the whole point of the wave function—is to define a motion for the particles, and in terms of this motion, pointers end up pointing and experiments end up having results, the kinds of results that it was the whole point of quantum mechanics to explain. So the connection to physical reality in the theory is via what we call the primitive ontology of the theory, in Bohmian mechanics the positions of the particles.

The wave function would seem to be part of the ontology. It is real in that sense. It's not subjective in Bohmian mechanics—it has a rather real role to play: it has to govern the motion of the particles. But it is not part of the primitive ontology. Bohmian mechanics is fundamentally about particles and their motions, not wave functions.

Here, for the record, are the equations of Bohmian mechanics. First, you have for the wave function the usual Schrödinger equation:

$$i\hbar\frac{\partial\psi}{\partial t} \;=\; H\psi, \tag{4.1}$$

---

[1] We use lowercase letters, such as $\mathbf{q}_1, \ldots, \mathbf{q}_N$, for generic position and configuration variables in quantum theory, reserving uppercase letters for the actual positions and configurations. It is interesting that in orthodox quantum theory one also has generic position variables as arguments of the wave function, even though there are no (unmeasured) actual positions. We find this rather odd.

with the usual Schrödinger Hamiltonian

$$H = -\sum_{k=1}^{N} \frac{\hbar^2}{2m_k} \nabla_k^2 + V.$$

Here $\nabla_k = (\partial/\partial x_k, \partial/\partial y_k, \partial/\partial z_k)$ is the position gradient for the kth particle, and $V = V(q)$ is a real-valued function of the configuration called the potential energy function.

The only thing Bohmian mechanics adds—in addition to the positions of the particles as actual variables in the theory to be taken seriously, not just talked about in connection with measurements—is an equation of motion for the positions:

$$\frac{d\mathbf{Q}_k}{dt} = \frac{\hbar}{m_k} \mathrm{Im}\, \frac{\psi^* \nabla_k \psi}{\psi^* \psi}\, (\mathbf{Q}_1, \ldots, \mathbf{Q}_N), \tag{4.2}$$

expressing the velocity of the kth particle, $k = 1, \ldots, N$, in terms of the wave function and the positions of all the particles. (Here Im means imaginary part: $\mathrm{Im}(z) = b$ for a complex number $z = a + ib$.) This equation, the new equation in Bohmian mechanics, is kind of obvious. It is more or less the first thing you would guess if you asked yourself, what is the simplest motion of the particles that could reasonably be defined in terms of the wave function? (However, it may not look so obvious.)

It might seem a bit pointless to have ψ^* in both the numerator and the denominator of (4.2), so that it cancels, leaving just $\mathrm{Im}(\nabla_k \psi/\psi)$ times a prefactor, which is the same as $\nabla_k S/m_k$, the more familiar way of writing the right-hand side—the velocity field in Bohmian mechanics. There are two good reasons for writing it in the apparently more complicated form.

In this way, the formula makes sense automatically for particles with spin, for which we would need wave functions with many components, instead of the simple single-component complex-scalar valued wave functions appropriate for systems of particles without spin (spin-0 particles). For example, the wave function for a system of N spin-1/2 particles has 2^N components: the value of ψ at a given configuration Q is given by 2^N complex numbers, not just one. In such a case, it is not clear what $\mathrm{Im}(\nabla_k \psi/\psi)$ could possibly mean. Why should the x-component of this, like the x-component of the velocity of a particle, be a scalar? What could even be meant by the ratio of two multicomponent objects?

But if you interpret the products involving ψs in the numerator and denominator in the formula as "spinor inner products," involving sums of products of components of ψ—and that's the natural way to understand such expressions, the natural product to form with such multicomponent wave functions—then the very same formula that is valid for particles without spin remains valid for particles with spin, providing an equation of motion for such particles that does exactly what you want it to do. It works perfectly. So you don't need to do anything extra in Bohmian mechanics to deal with spin. That's one reason for writing the equation as above.

The other reason is that the denominator, $\psi^*\psi$, is the familiar quantum probability density ρ, and the numerator the quantum probability current $\boldsymbol{J}_k$. Thus the right-hand side is $\boldsymbol{J}_k/\rho$, a fairly obvious thing to guess for a velocity. Because the velocity is $\boldsymbol{J}_k/\rho$, the $|\psi|^2$-probabilities play the role they do in Bohmian mechanics.

As a consequence of this role, the usual quantum randomness emerges. One obtains the *quantum equilibrium hypothesis,* that whenever a system has wave function ψ, its configuration is random, with distribution given by $|\psi|^2$. Exactly what this means and how this comes about is a long and controversial story, which we do not go into here. Using the quantum equilibrium hypothesis, one can establish the empirical equivalence between Bohmian mechanics and orthodox quantum theory, including the emergence of operators as "observables" and the collapse of the wave packet (Dürr, Goldstein, and Zanghì 1992, 2004).

3 The Wave Function of a Subsystem

A crucial ingredient in the extraction of the implications of Bohmian mechanics is the notion of the wave function of a subsystem of a Bohmian universe, a universe of particles governed by the equations of Bohmian mechanics, defining a motion choreographed by the *wave function of the universe* Ψ.

In almost all applications of quantum mechanics, the wave function of a subsystem is what we are concerned with, not the wave function of the universe. The latter, after all, must be rather elusive. Most physicists don't deal with the universe as a whole. They deal with subsystems more or less all the time: a hydrogen atom, particles going through Stern-Gerlach magnets, a Bose-Einstein condensate, and so on. Yet from a fundamental point of view, the only genuine Bohmian system in a Bohmian universe—the only system you can be sure is Bohmian—is the universe itself, in its entirety. It cannot be an immediate consequence of this fact that subsystems of a Bohmian universe are themselves Bohmian, with the motion of their particles governed by wave functions in the Bohmian way.

That is, one can't simply demand of subsystems of a Bohmian universe that they be Bohmian systems in their own right. The behavior of the parts of a big system are already determined by the behavior of the whole. What you have for the whole is the wave function Ψ of the universe, together with its configuration Q. That's your data. That's what is objective in a Bohmian universe. The wave function of a subsystem, if it exists at all, must be definable in terms of that data.

Corresponding to a subsystem of the universe is a splitting $Q = (Q_{sys}, Q_{env}) =$ (X, Y) of its configuration Q into the configuration $Q_{sys} = X$ of the subsystem, the "x-system," formed from the positions of the particles of the subsystem, and the configuration $Q_{env} = Y$ of the environment of the subsystem—the configuration of everything else. So the data in terms of which the wave function of a subsystem must be defined are the universal wave function $\Psi(q) = \Psi(x,y)$ and the actual configurations X of the subsystem and Y of its environment.

The first guess people make about what the wave function $\psi(x)$ of the x-subsystem should be usually turns out to be wrong. The right guess, and the natural thing to do, is to define the wave function of a subsystem in this way: remembering that the wave function of a subsystem should be a function on its configuration space (a function, that is, of x alone), you take the universal wave function $\Psi(x, y)$ and plug the actual configuration Y of the environment into the second slot to obtain a function of x,

$$\psi(x) = \Psi(x, Y). \tag{4.3}$$

If you think about it, you see that this is exactly the right definition. The situation is simplest for spin-0 particles, which we henceforth assume. First, it is easy to see that the velocity that the configuration X of the subsystem inherits from the motion of the configuration Q can be expressed in terms of this ψ in the usual Bohmian way. In other words, if $dQ/dt = v^{\Psi}(X, Y)$ then $dX/dt = v^{\psi}(X)$ for $\psi(x) =$ $\Psi(x, Y)$.

However, the evolution law for the wave function of the subsystem need not be Bohmian. Explicitly putting in the time dependence, we have for the wave function of the x-system at time t:

$$\psi_t(x) = \Psi_t(x, Y_t).$$

Thus the wave function of a subsystem has an interesting time dependence. Time appears in two places here. The wave function of the universe depends on t because it evolves according to Schrödinger's equation. The configuration of the environment Y also evolves and depends on t as part of the evolving configuration of the universe $Q_t = (X_t, Y_t)$.

This suggest a rich variety of ways the wave function of a subsystem might behave in time. Everyone readily believes, and it is in fact the case, that the wave function of a subsystem evolves just as it should for a Bohmian system, according to Schrödinger's equation for the subsystem, when the subsystem is suitably decoupled from its environment. And it is actually rather easy to see that the wave function of a subsystem collapses according to the usual textbook rules with the usual textbook probabilities in the usual measurement situations. The wave function of the x-system thus collapses in just the way wave functions in quantum mechanics are supposed to collapse. This follows more or less directly from standard quantum measurement theory together with the definition of the wave function of the x-system and the quantum equilibrium hypothesis (Dürr, Goldstein, and Zanghì 1992).

It is a sociological fact, for whatever reason, that even very talented mathematical physicists have a lot of trouble accepting that the wave function of a subsystem collapses as claimed. We guess that is because people know that collapse in quantum mechanics is supposed to be some really problematic, difficult issue, so they think it can't be easy for Bohmian mechanics either. But it is easy for Bohmian mechanics. What everyone is happy to take for granted—that the wave function of a subsystem will evolve according to Schrödinger's equation in the appropriate situations—is taken for granted presumably because nobody says there is a problem getting wave functions to obey Schrödinger's equation. Collapse is the problem. But understanding why the wave function of a subsystem does evolve according to Schrödinger's equation when the subsystem is suitably decoupled from its environment is a bit tricky. Nonetheless it is true, though we shall not go into any details here, see Dürr, Goldstein, and Zanghì (1992).

The main point we wish to have conveyed in this section is that for a Bohmian universe, the wave function of a subsystem of that universe—defined in terms of the wave function of the universe and additional resources available to Bohmian mechanics and absent in orthodox quantum theory (namely, the actual configuration of the environment of the subsystem)—behaves exactly the way wave functions in orthodox quantum theory are supposed to behave.

4 The Wave Function as Nomological

The main thing we want to discuss here is the status of the wave function: what kind of thing it is. We want to suggest that one should think about the possibility that it is nomological, nomic—more in the nature of a law than a concrete physical reality.

Thoughts in this direction might arise when one considers the unusual way Bohmian mechanics is formulated and the unusual sort of behavior that the wave function undergoes in Bohmian mechanics. The wave function of course affects the behavior of the configuration, that is, of the particles. This is expressed by the guiding equation (4.2), which in more compact form can be written

$$dQ/dt = v^{\Psi}(Q). \qquad (4.4)$$

But in Bohmian mechanics there's no back action, no effect in the other direction, of the configuration on the wave function, which evolves autonomously via Schrödinger's equation (4.1), in which the actual configuration Q does not appear. Indeed, the actual configuration could not appear in Schrödinger's equation because this equation is also in orthodox quantum theory, and in orthodox quantum theory there is no actual position or configuration. That's one point.

A second point is that for a multiparticle system the wave function $\psi(q) = \psi(\mathbf{q}_1, \ldots, \mathbf{q}_N)$ is not a weird field on physical space, it's a weird field on configuration space, the set of all hypothetical configurations of the system. For a system of more than one particle, that space is not physical space. What kind of thing is this field on that space?[2]

The fact that Bohmian mechanics requires that one take such an unfamiliar sort of entity seriously bothers a lot of people. It does not bother us all that much, but it does seem like a significant piece of information. What it suggests to us is that you should think of the wave function as describing a law, not as some sort of concrete physical reality. After all (4.4) is an equation of motion, a law of motion, and the whole point of the wave function here is to provide us with the law, that is, with the right-hand side of this equation.

We said that rather cavalierly. There are lots of problems with saying it at this point. Before going into the problems, let us make a comparison with a familiar situation with which nobody seems to have much of a problem at all: classical Hamiltonian dynamics.

4.1 Comparison of ψ with the Classical Hamiltonian H

The wave function is strange because it lives on configuration space, for an N-particle system a space of dimension $3N$. Well, there's a space in the classical mechanics of an N-particle system that has twice that dimension, its phase space, of dimension $6N$. On that space there's a function, the Hamiltonian $H = H(q, p) = H(\mathcal{X})$ of the system, and to define the equations of motion of classical

---

[2] The sort of physical reality to which the wave function corresponds is even more abstract than what we have conveyed so far. That's because the wave function, in both orthodox quantum theory and Bohmian mechanics, is merely a convenient representative of the more physical "quantum state." Two wave functions such that one is a (nonzero) scalar multiple of the other represent the same quantum state and are regarded as physically equivalent. Thus, the quantum state is not even a field at all, but an equivalence class of fields. It is worth noting that equivalent wave functions define the same velocity (4.2). They also define, with suitable normalization, the same $|\psi|^2$-probabilities. Moreover, for the treatment of identical particles such as electrons in Bohmian mechanics, it is best to regard them as unlabeled, so that the configuration space of N such particles is not a high-dimensional version of a familiar space, like $\mathbb{R}^{3N}$, but is instead the unfamiliar high-dimensional space $^N\mathbb{R}^3$ of N-point subsets of $\mathbb{R}^3$. This space has a nontrivial topology, which naturally leads to the possibilities of bosons and fermions—and in two dimensions anyons as well (Dürr et al. 2006). As a fundamental space it is odd, but not as a configuration space.

mechanics you put H on the right-hand side of the equations of motion after suitably taking derivatives. We have never heard anyone complaining about classical mechanics because it invokes a weird field on phase space and asking about what kind of thing that is. No one has any problem with that. Everyone knows that the Hamiltonian is just a convenient device in terms of which the equations of motion can be nicely expressed.

We suggest that you should regard the wave function in exactly the same way. If you want to have a sharper analogy, you can think not of ψ itself but of something like $\log \psi(q)$ as corresponding to the Hamiltonian $H(\mathcal{X})$. The reason we suggest this is because the velocity in Bohmian mechanics is proportional to the imaginary part of $\nabla\psi/\psi$ for a scalar wave function, a sort gradient of the log of ψ, some sort of derivative, der, of $\log \psi(q)$, so that (4.2) can be regarded as of the form

$$dQ/dt = der \, (\log \psi).$$

Similarly in classical mechanics we have an evolution equation of the form

$$d\mathcal{X}/dt = der \, H,$$

where $der \, H$ is a suitable derivative of the Hamiltonian. (This is a compact way of writing the familiar Hamiltonian equations $d\mathbf{q}_k/dt = \partial H/\partial \mathbf{p}_k$, $d\mathbf{p}_k/dt = -\partial H/\partial \mathbf{q}_k$.)

It is also true that both $\log \psi$ and H are normally regarded as defined only up to an additive constant. When you add a constant to H, it doesn't change the equations of motion. If you multiply the wave function by a scalar—which amounts to adding a constant to its log—the new wave function is generally regarded as physically equivalent to the original one. This is so for orthodox quantum theory because if wave functions ψ and ψ' are related by $\psi' = c\psi$ (with the constant $c \neq 0$), then the predictions of quantum theory for the probabilities of measurement results when a system is in state ψ are exactly the same as those for a system in state ψ'. In Bohmian mechanics, the new wave function defines the same velocity for the configuration, the same equations of motion, as the original one.

Moreover, with suitably "normalized" choices of $\psi(q)$ and $H(\mathcal{X})$, corresponding to appropriate choices of the constants, one associates rather similar probability formulas. In classical statistical mechanics there are the Boltzmann-Gibbs probabilities, given by $e^{-H/kT}$ when H has been suitably normalized, where k is Boltzmann's constant, and T is the temperature. One thus has that

$$\log Prob \propto - H.$$

In quantum mechanics or Bohmian mechanics, with $|\psi|^2$-probabilities, one has that

$$\log Prob \propto \log |\psi|.$$

(You probably should not take this last point about analogous probabilities too seriously. It's presumably just an accident that the analogy seems to extend this far.)

4.2 ψ versus Ψ

There are, however, problems with regarding the wave function as nomological. Laws aren't supposed to be dynamical objects, they aren't supposed to change with time, but the wave function of a system typically does. Laws are not supposed to be things that we can control—we're not God. But the wave function is often an initial condition for a quantum system. Often, in fact, we prepare a system in a certain quantum state, that is, with a certain wave function. In this sense we can control the wave function of a system. But we do not control a law of nature. This makes it a bit difficult to regard the wave function as nomological.

With regard to this difficulty, it is important to recognize that there's only one wave function we should be worrying about, the fundamental one, the wave function Ψ of the universe. In Bohmian mechanics, the wave function ψ of a subsystem of the universe is defined in terms of the universal wave function Ψ. Thus, to the extent that we can grasp the nature of the universal wave function, we should understand as well, by direct analysis, the nature of the objects that are defined in terms of it; in particular, we should have no further fundamental question about the nature of the wave function of a subsystem of the universe. So we focus on the former.

4.3 The Universal Level

When we consider, instead of the wave function of a typical subsystem, the wave function Ψ of the universe itself, the situation is rather dramatically transformed. Ψ is not controllable. It is what it is! It may well not be dynamic either. There may well be no t in Ψ.

The fundamental equation for the wave function of the universe in canonical quantum cosmology is the Wheeler–DeWitt equation (DeWitt 1967),

$$\mathcal{H}\Psi = 0,$$

for a wave function $\Psi(q)$ of the universe, where q refers to three-geometries and whatever other stuff is involved, all of which correspond to structures on a three-dimensional space. In this equation, $\mathcal{H}$ is a sort of generalized Laplacian, a cosmological version of a Schrödinger Hamiltonian H. Like a typical H, it involves nothing like an explicit time dependence. But unlike Schrödinger's equation, the Wheeler–DeWitt equation has on one side, instead of a time derivative of Ψ, simply 0. Its natural solutions are thus time-independent, and these are the solutions of the Wheeler–DeWitt equation that are relevant in quantum cosmology.

This is, in fact, the *problem of time* in quantum cosmology. We live in a world where things change. If the basic object in the world is a timeless wave function, how does change come about? Much has been written about this problem of time, and a great many answers have been proposed. What we want to emphasize here is that from a Bohmian perspective the timelessness of Ψ is not a problem. Rather, it is just what the doctor ordered.

The fundamental role of the wave function in Bohmian mechanics is to govern the motion of something else. Change fundamentally occurs in Bohmian mechanics not so much because the wave function changes but because the thing Q it's governing does, according to a law

$$dQ/dt = v^{\Psi}(Q) \qquad (4.5)$$

determined by the wave function. The problem of time vanishes entirely from a Bohmian point of view. This is just what the doctor ordered because laws are not supposed to change with time, so we don't want the fundamental wave function to change with time. It's good that it does not change with time.

There may be another good thing about the wave function of the universe: it may be unique. Of course, along with being uncontrollable, a timeless wave function of our actual universe would be the one wave function that it is. But we mean more than that. Although the Wheeler–DeWitt equation presumably has a great many solutions Ψ, when supplemented with additional natural conditions (for example the Hartle-Hawking boundary condition; Hartle and Hawking 1983), the solution may become unique. Such uniqueness fits nicely with the conception of the wave function as law.

4.4 Schrödinger's Equation as Phenomenological (Emergent)

Now we can well imagine someone saying, "Okay, fine, in this Bohmian theory for the universe stuff changes—particles move, the gravitational field changes, the gravitational metric evolves, whatever. But we know that the most important equation in quantum mechanics, and one of the most important equations in our quantum world, is the time-dependent Schrödinger equation, describing wave functions that themselves change with time. Where does that come from in a theory in which the only fundamental wave function that you have is the timeless wave function Ψ?"

That question has already been answered here, in the last paragraph of section 3. If you have a wave function of the universe obeying Schrödinger's equation, then in suitable situations, those in which a subsystem is suitably decoupled from its environment (and the Hamiltonian $\mathcal{H}$ is of the appropriate form), the wave function $\psi_t(x) = \Psi(x, Y_t)$ of the subsystem will evolve according to Schrödinger's equation for that subsystem. For Ψ not depending on time, the wave function of the subsystem inherits its time dependence from that of the configuration Y_t of the environment. The crucial point here is that a solution Ψ_t of Schrödinger's equation can be time-independent. These are the solutions Ψ_t that involve the same wave function Ψ for all t, the solutions for which $\partial\Psi_t/\partial t$ is 0 for all t, corresponding precisely to solutions of the Wheeler–DeWitt equation.

In this situation, the time-dependent Schrödinger evolution (4.1) is not fundamental. Rather, it is emergent and phenomenological, arising—as part of a good approximation for the behavior of suitable subsystems—from a Bohmian dynamics (4.5) for the universe given in terms of a suitable wave function of the universe, one that obeys the Wheeler–DeWitt equation.

Even this time-independent equation might not be fundamental—that it appears to be might be an illusion. What we have in mind is this. We have a law of motion involving a vector field v^Ψ (the right-hand side of a first-order equation of motion), a vector field that can be expressed in terms of Ψ. If Ψ is a nice sort of wave function, it might obey all sorts of nice equations, for example the Wheeler–DeWitt equation or something similar. From a fundamental point of view, it might be a complete accident that Ψ obeys such an equation. It might just happen to do so. The fact that the equation is satisfied might have nothing to do with why the fundamental dynamics is of the form (4.5). But as long as Ψ does satisfy the equation, by accident or not, all the consequences of satisfying it follow.

So it could turn out, at the end of the day, that what we take to be the fundamental equation of quantum theory, Schrödinger's equation, is not at all fundamental for quantum theory, but is an emergent and accidental equation.

4.4.1 Two Transitions

We want to focus a bit on the change of perspective that occurs when we make the transition from orthodox quantum theory (OQT), which seems to involve only the wave function ψ, to (conventional) Bohmian mechanics (BM), which is usually regarded as involving two types of physical entities, wave functions ψ and the positions of particles, forming a configuration Q, to universal Bohmian mechanics (UBM), where the wave function Ψ is taken out of the category of concrete physical reality and into that of law, so you have just Q as describing elements of physical reality:

$$\underset{\psi}{OQT} \xrightarrow{BM} \underset{(\psi, Q)}{} \xrightarrow{UBM} \underset{Q}{} \; .$$

You start with just ψ, you end with just Q.

Our original question, about the wave function—what kind of thing is that?—is rather dramatically transformed when we make this transition to the universal level, because we are then asking about a very different object, not about a wave function of a subsystem of the universe but about the universal wave function:

$$? \; \overset{?}{\underset{?}{\psi}} \; ? \; \to \; ? \; \overset{?}{\underset{?}{\Psi}} \; ?,$$

which is actually, we are supposing, just a way of representing the law of motion. So now we may ask, does any kind of question about Ψ remain?

Here's one question: why should the motion be of the form (4.5), involving Ψ in the way that it does? Why should the law of motion governing the behavior of the constituents of the universe be of such a form that there is a wave function in terms of which the motion can be compactly expressed? We think that's a good question. Of course, we have no definitive answer to it. But an answer to this question would provide us with a deep understanding of why our world is quantum mechanical.

The view that the wave function is nomological has another implication worth considering. This is connected with the question of how we ever come to know what the wave function of a system is. There must be some algorithm that we use. We do not directly see wave functions. What we see (more directly) are particles, at least from a Bohmian perspective. We should read off from the state of the primitive ontology, whatever it may be, what the relevant wave function is. There should be some algorithm connecting the state of the primitive ontology, for Bohmian mechanics the relevant configuration $Q(t)$ over, say, some suitable time interval, with the relevant wave function.

Now let's go to the universal level. You might think there should be some algorithm that we can use to read off the universal wave function from the state of the primitive ontology of the universe, whatever that may be. But in fact, we kind of doubt there is any such algorithm. As far as we know, nobody has proposed any such algorithm. From the point of view of the wave function Ψ being nomological, you would not expect there to be any such algorithm. That's because if the wave is nomological, specifying the wave function amounts to specifying the theory. You would not expect there to be an algorithm for theory formation.

4.5 Nomological versus Nonnomological

Now, we can imagine—and in fact we are quite sure—that many physicists would respond to the question about whether the wave function is fundamentally nomological with a big "Who cares? What difference does it make?"

We think it does matter. Being nomological has important implications. Laws should be simple. If we believe that the wave function of the universe is nomological, this belief should affect our expectations for the development of physics. We should expect somehow to arrive at physics in which the universal wave function involved in that physics is in some sense simple—while presumably having a variety of other nice features as well.

Simplicity itself is sort of complicated. There are a number of varieties of simplicity. For example, the universal wave function could be simple in the sense that it has a simple functional form—that it's a simple function of its arguments. That is one possibility. Another, quite different one is that it could be a more or less unique solution to a simple equation. Or, a similar kind of thing, it could more or less uniquely satisfy some compelling principle, maybe a symmetry principle.[3]

4.6 Relativistic Bohmian Theory

We want to say a bit about Lorentz invariance and a problem that arises in connection with it. It is widely said—and it's natural to think—that you cannot have a Lorentz-invariant Bohmian theory. That's basically because of the crucial role played in such a theory by the configuration of the system: the positions of its particles—or the detailed description of the primitive ontology of the theory, whatever that may be—*at a given time.*

Now you can also consider configurations determined, not by a t = constant hypersurface but by a general space-like hypersurface. For example, for a particle ontology, the configuration corresponding to such a surface would be given by the space-time points on the surface at which the world-lines of the particles cross the surface, for N particles, N points. So if in fact you had somehow at your disposal a Lorentz-invariant foliation of space-time into space-like hypersurfaces, you could play a Bohmian game and define a Bohm-type dynamics for the evolution of

---

[3] For example, Goldstein and Teufel (2001) explored the possibility that a symmetry principle expressing a sort of quasi four-diffeomorphism invariance would imply an evolution of three-geometries governed by a universal wave function. In technical terms, we demanded that the vector field on super-space defining the relevant motion form a representation of the "Dirac algebra" (Dirac 1964), a sort of algebra, sort of corresponding to four-diffeomorphism invariance. That puts very strong constraints on the theory. In fact the constraints are so strong that it seems the only possibilities correspond to classical general relativity, with nothing genuinely quantum mechanical arising. That's for pure gravity. It is not clear what would happen if matter degrees of freedom were included in the analysis. So one could always hope that if matter were included in the story and you played a similar game, you would thereby end up with quantum mechanics as the only (reasonable) possibility. But that is highly speculative—only a hope and a prayer.

configurations defined in terms of that foliation. In this way, one could obtain a Lorentz-invariant Bohmian theory (Dürr et al. 1999).

To actually have such a thing, the best possibility is perhaps the following. You have a Lorentz-invariant rule for defining in terms of the universal wave function a *fol* of space-time, a covariant map *fol* from wave functions to foliations:

$$(\text{Lorentz}) \text{ covariant map } \Psi \xrightarrow{\;fol\;} \mathscr{F} = \mathscr{F}\,(\Psi).$$

(For the map to be *covariant* means that the diagram

$$\Lambda_g \downarrow \;\;\; \begin{array}{ccc} \Psi & \xrightarrow{\;fol\;} & \mathscr{F} \\ & & \downarrow g \\ \Psi_g & \xrightarrow{\;fol\;} & \mathscr{F}_g \end{array}$$

is commutative. Here, g is any Lorentz transformation, on the right acting naturally on the foliation by moving the points on any leaf of the foliation, and hence the leaves themselves and the foliation itself, around according to g, and Λ_g is the action of g on wave functions, given by a representation Λ of the Lorentz group.)

Lorentz-invariant Bohmian theories formed in this way, by using such a covariant foliation map, have the virtue of being seriously Lorentz-invariant. The point here is that any theory can be made Lorentz-invariant in a trivial nonserious way by introducing suitable additional space-time structure beyond the Lorentz metric. The question then arises as to what kinds of structure are unproblematic. James Anderson (1967) has addressed this question by distinguishing between absolute and dynamic structures and identifying the serious Lorentz invariance of a theory with the nonexistence in the theory of any additional absolute structures.

What exactly these are is not terribly relevant for our purposes here. That's because, for the sort of theory proposed here, what seems to be additional space-time structure, namely, the foliation, is not an additional structure at all beyond the wave function. To the extent that the wave function is a legitimate structure for a Lorentz-invariant theory—and this is generally assumed to be the case—so are covariant objects defined solely in terms of the wave function.

Here are some examples of possibilities for covariant foliations. You could form a typical quantum expectation in the Heisenberg picture, involving the universal wave function and some sort of operator-valued Fermi field $\psi(x)$. The simplest such object is perhaps

$$j_\mu(x) = \langle \Psi | \bar{\psi}(x)\gamma_\mu\,\psi(x) | \Psi \rangle,$$

involving the Dirac matrices γ_μ, defining a time-like vector field on space-time. You could also put suitable products in the middle to form tensors of various ranks. Ward Struyve (see Dürr et al. in preparation) has suggested using the stress energy tensor

$$t_{\mu\nu}(x)=\langle\Psi|T_{\mu\nu}(x)|\Psi\rangle$$

and integrating that over space-like hypersurfaces to obtain a time-like vector (that in fact does not depend on the choice of surface). There are a variety of such proposals for extracting from the wave function a vector field on space-time in a covariant manner. A vector field on space-time is just the sort of thing that could define a foliation, namely, into hypersurfaces orthogonal to that vector field, so that we have the following scheme for a map *fol*:

$$\Psi \rightarrow j_\mu \rightsquigarrow \mathscr{F}.$$

Struyve's proposal works as is, but for other proposals you would have to do a lot of massaging to get the scheme to work. To define a foliation, the vector field would have to be what is called "in involution." That can be achieved, but in doing so you would like the resulting vector field to remain time-like (so that the corresponding foliation would be into space-like hypersurfaces), and that is certainly not automatic.

The bottom line is that there is lots of structure in the universal wave function—certainly enough structure to typically permit the specification of a covariant rule for a foliation.

4.7 Wave Function as Nomological and Symmetry

There is a problem: there is a conflict between the wave function being nomological and symmetry demands. The problem arises from the difference between having an action of the Lorentz group G (or whatever other symmetry group we have in mind) on the Hilbert space $\mathscr{H}$ of wave functions (or on a suitable subset of $\mathscr{H}$, the domain of the foliation map)—which is more or less all that is usually required for a Lorentz-invariant theory—and having the trivial action, always carrying Ψ to itself: the difference between an action of G on $\mathscr{H}$ and the G-invariance of Ψ. If the universal wave function represents the law, then that wave function itself, like an invariant law, should be G-invariant. (Actually, any change of the wave function that leaves the associated velocity vector field alone would be fine, for example. multiplication by a constant scalar, but we shall for simplicity ignore this possibility here.)

It is not hard to see that that's incompatible with the covariance of the foliation map. No foliation can be Lorentz-invariant, because there is always some Lorentz transformation that will tilt some of its leaves, at least somewhere. But if Ψ is g-invariant so must be any foliation associated with Ψ in a covariant manner. Thus a Lorentz-invariant wave function Ψ cannot be covariantly associated with a foliation.

This is in sharp contrast with the situation for a generic wave function of the universe, one that is in no way special. Such a wave function will not be symmetric, and there is no obstacle to its being in the domain of a covariant foliation map. But if the universal wave function is nomological, it is not generic, and it must be too symmetric to permit the existence of a covariant foliation map.

4.7.1 Possible Resolutions

There seems to be a conflict between (1) having a Bohmian quantum theory, (2) the universal wave function for that theory being nomological, and (3) fundamental Lorentz invariance. Something has to give, it would seem. From a Bohmian point of view, the thing that gives wouldn't be the Bohmian part.

Here are some possible resolutions that would allow us to continue to regard the wave function as nomological. You could abandon fundamental Lorentz invariance, as many have suggested. Another possibility is to make use of a Lorentz-invariant foliation—one not determined by the wave function but defined in terms of additional dynamical structure beyond the wave function, for example, some suitable time-like vector field definable from the primitive ontology or perhaps transcending the primitive ontology, or something like a "time function" in general relativity, defined in terms of the gravitational metric and stuff in space-time. Or, the most likely possibility: something no one has thought of yet.

Of course, there is the possibility that we will have to abandon our attempt to regard the wave function as nomological. Many (for example Travis Norsen), would then insist that the wave function be eliminated in favor of something like exclusively local beables (Norsen 2010). That's not how we feel. If it should turn out that the wave function cannot be regarded as nomological—because it's too complicated or whatever—our reaction would probaby be: "Okay, that's just the way it is. It's not nomological but something different."

We think, in fact, that if someone gave us a Bohmian kind of theory, involving a complicated collection of exclusively local beables, and then someone else pointed out to us that the complicated local beables can be repackaged into a simple mathematical object of a nonlocal character—like a wave function on configuration space—our reaction would likely be that we would prefer to regard the wave function in that simpler though more unfamiliar way, just because of its mathematical simplicity.

4.8 ψ as Quasi-Nomological

Suppose we accept that the universal wave function Ψ is nomological. What, then, about the status of the wave function ψ of subsystems of the universe—the wave functions with which we're normally concerned in applications of quantum theory? We have several responses.

Our first response is this: you can decide for yourself. We are assuming the status of Ψ is clear. The status of the primitive ontology is certainly clearer still. Therefore, because ψ is defined in terms of Ψ and the primitive ontology (specifically, the configuration Y of the environment), the status of ψ must follow from an analysis of its definition.

We do not insist that everyone would agree on the conclusion of such an analysis. It may well be that different philosophical prejudices will lead to different conclusions here. Our point is rather that once the status of the wave function of the universe has been settled, the question about the status of ψ is rather secondary—something about which one might well feel no need to worry.

Be that as it may, we would like to regard ψ as quasi-nomological. By this we mean that although there are serious obstacles to regarding the wave function of a subsystem as fully nomological, ψ does have a nomological aspect in that it seems more like an entity that is relevant to the behavior of concrete physical reality (the primitive ontology) and not so much like a concrete physical reality itself.

We can say more. The law governing the behavior of the primitive ontology of the universe naturally implies a relationship between the behavior of a subsystem and the configuration of its environment. It follows from its definition (4.3) that the wave function of the subsystem captures that aspect of the environment that expresses this relationship—that component of the universal law that is relevant to the situation at hand, corresponding to the configuration of the environment.

5 The Status of the Wave Function in Quantum Theory

Let's return to the possibilities for the wave function, mentioned in section 1. It could be nothing. Though not exactly nothing, it could be merely subjective or epistemic, representing our information about a system. Or it could be something objective. If it is objective, it could be material or quasi-material, or it could be nomological, or at least quasi-nomological.

Rather than deciding in absolute terms which of these possibilities is correct or most plausible—concerning which our opinion should be quite clear—we conclude by stressing that one's answer to this question should depend on one's preferred version of quantum theory. Here are some examples.

- In orthodox quantum theory, the wave function is quasi-nomological. It governs the results of quantum "measurements"—it provides statistical relationships between certain macroscopic variables.
- In Everett, the wave function is quasi-material. After all, there it's all there is. In Everett there's only the wave function. (We say "quasi-material" here instead of plain material because in Everett the connection between the wave function and our familiar material reality is not at all straightforward. In fact, for some Everettians part of the appeal of their approach is the extensive conceptual functional analysis that it requires, see, e.g., Wallace 2003.)
- In Bohmian mechanics as we understand it, as well as in decoherent or consistent histories and in causal set theory, the wave function is either nomological or quasi-nomological. In these theories, the wave function governs the behavior of something else, something more concretely physical.
- However, in David Albert's (1996) version of Bohmian mechanics, in which what we call configuration space is in fact a very high-dimensional *physical* space, on which the wave function lives as a physical field, the wave function is material or quasi-material.
- In Ghirardi-Rimini-Weber theory (Ghirardi, Rimini, and Weber 1986) or continuous spontaneous localization (Ghirardi, Pearle, and Rimini 1990), the wave function is quasi-material, because it either is everything or at least determines everything.
- In the quantum information approach to quantum theory, the wave function is quasi-subjective—"quasi" because quantum information theorists differ as to how subjective it is.

This list indicates that if you want to grasp the status of the wave function in quantum theory, you need to know exactly what quantum theory says. If you are not clear about quantum theory, you should not be worrying about its wave function.

We have suggested seriously considering the possibility that the wave function is nomological. One psychological obstacle to doing so is that it seems to be an important feature of wave functions that they are variable, and this variability—from system to system and not just over time—leads to the varieties of different behaviors that are to be explained by quantum theory. But the behavior of the primitive ontology of a Bohmian theory, and all of the empirical consequences of the theory, depend on the universal wave function only via the one such wave function that exists in our world and not on the various other universal wave functions that there might have been. The variability we see in wave functions is that of wave functions of subsystems of the universe. This variability originates in that of the environment Y of the subsystem as well as that of the choice of subsystem itself. So this variability does not conflict with regarding the wave function as fundamentally nomological, but is explained by it.

References

Albert, D. Z., Elementary Quantum Metaphysics. In J. Cushing, A. Fine, and S. Goldstein (eds.), *Bohmian Mechanics and Quantum Theory: An Appraisal,* Dordrecht: Kluwer (1996).

Allori, V., Goldstein, S., Tumulka R., and Zanghì, N., On the Common Structure of Bohmian Mechanics and the Ghirardi-Rimini-Weber Theory. *British Journal for the Philosophy of Science* 59:353–89 (2008). arXiv:quant-ph/0603027.

Anderson, J. L., *Principles of Relativity Physics.* New York: Academic Press (1967).

Bell, J. S., On the Problem of Hidden Variables in Quantum Mechanics. *Reviews of Modern Physics* 38:447–452 (1966). Reprinted in J. S. Bell, *Speakable and Unspeakable in Quantum Mechanics,* Cambridge: Cambridge University Press (1987), chapter 1.

Bohm, D., A Suggested Interpretation of the Quantum Theory in Terms of "Hidden" Variables, I and II. *Physical Review* 85:166–93 (1952).

DeWitt, B. S., Quantum Theory of Gravity. I. The Canonical Theory. *Physical Review* 160:1113–48 (1967).

Dirac, P. A. M., *Lectures on Quantum Mechanics,* New York: Yeshiva University (1964).

Dürr, D., Goldstein, S., Münch-Berndl, K., and Zanghì, N., Hypersurface Bohm-Dirac models. *Physical Review A* 60:2729–36 (1999). arXiv:quant-ph/9801070.

Dürr, D., Goldstein, S., Struyve, W, Norsen, T., and Zanghì, N., Can Bohmian Mechanics Be Made Relativistic? In preparation.

Dürr, D., Goldstein, S., Taylor, J., Tumulka, R., and Zanghì, N., Topological Factors Derived from Bohmian Mechanics. *Annales Henri Poincaré* 7:79–807 (2006). arxiv:quant-ph/0601076.

Dürr, D., Goldstein, S., and Zanghì, N., Quantum Equilibrium and the Origin of Absolute Uncertainty. *Journal of Statistical Physics* 67:843–907 (1992). arXiv:quant-ph/0308039.

Dürr, D., Goldstein, S., and Zanghì, N., Quantum Equilibrium and the Role of Operators as Observables in Quantum Theory. *Journal of Statistical Physics* 116:959–1055 (2004). arxiv:quant-ph/0308038.

Everett, H., Relative State Formulation of Quantum Mechanics. *Reviews of Modern Physics* 29:454–62 (1957).

Ghirardi, G. C., Pearle, P., and Rimini, A., Markov Processes in Hilbert Space and Continuous Spontaneous Localization of Systems of Identical Particles. *Physical Review A* 48:78–89 (1990).

Ghirardi, G. C., Rimini, A., and Weber, T., Unified Dynamics for Microscopic and Macroscopic Systems. *Physical Review D* 34:470–91 (1986).

Goldstein, S., Quantum Theory without Observers. *Physics Today,* Part One: March 1998, 42–46. Part Two: April 1998, 38–42.

Goldstein, S., Bohmian Mechanics. *Stanford Encyclopedia of Philosophy* (2001), available online at http://plato.stanford.edu/entries/qm-bohm/.

Goldstein, S., and Teufel, S., Quantum Spacetime without Observers: Ontological Clarity and the Conceptual Foundations of Quantum Gravity. In C. Callender and N. Huggett (eds.), *Physics Meets Philosophy at the Planck Scale,* Cambridge: Cambridge University Press (2001).

Hartle, J., and Hawking, S., Wave Function of the Universe. *Physical Review D* 28:2960–75 (1983).

Norsen, T., The Theory of (Exclusively) Local Beables. *Foundations of Physics* 40:1858–84 (2010).

Wallace, D., Everett and Structure. *Studies in History and Philosophy of Modern Physics* 34:87–105 (2003).

5

Dimension and Illusion[*]

PETER J. LEWIS

> And even as we, who are now in Space, look down on Flatland and
> see the insides of all things, so of a certainty there is yet above us
> some higher, purer region, whither though dost surely purpose
> to lead me—O my Priest, Philosopher, and Friend—some yet
> more spacious Space, some more dimensionable Dimensionality,
> from the vantage-ground of which we shall look down together
> upon the revealed insides of Solid things, and where thine own
> intestines, and those of thy kindred Spheres, will lie exposed to
> the view of the poor wandering exile from Flatland, to whom so
> much has already been vouchsafed. (Abbott 1952, p. 88)

Could the three-dimensionality of the world be an illusion? That is, could the
world *appear* to have a different number of spatial dimensions than it *actually*
has? If the question concerns mere possibility, then the answer is clearly yes—
there are many ways this might be. The most straightforward, perhaps, is the
scenario explored in *Flatland* (Abbott 1952). Consider a race of creatures whose
movements and sensory experiences are confined to a plane within a three-
dimensional world. Such a world appears to its inhabitants to be two-dimensional. However, the illusory nature of this appearance can be revealed to them
by interaction with an ordinary three-dimensional being like us, via such phenomena as appearing "from nowhere" within a locked room or describing the
current state of a Flatlander's intestines. Similarly, we can imagine a race of
creatures whose movements and sensory experiences are confined to a three-
dimensional space within a four-dimensional world. Such a world appears to
its inhabitants to be three-dimensional, and again the illusory nature of this
appearance could be revealed by interaction with four-dimensional beings via
the kinds of phenomena just mentioned.

So the three-dimensionality of the world *could* be an illusion. But is there
any reason to think that it *is* an illusion? Certainly creatures don't materialize
inside locked rooms, and if there are beings who can describe the current state

---

[*] I thank Alyssa Ney and David Albert for very helpful comments on an earlier draft of this chapter.

of our intestines to us, they don't do so by hovering over us in a fourth dimension (unfortunately). However, modern physics has led to various claims that the dimensionality of the world is greater than three. First, special relativity arguably entails that the time dimension is not distinct from the three spatial dimensions, and hence the world is four-dimensional. Although there are interesting questions here about the sense in which the time dimension can be treated as "space-like," I restrict attention for present purposes to spatial dimensions, excluding time. Second, string theory postulates that there are 9 (or perhaps 10) spatial dimensions, where all but three are "curled up" so small that they are irrelevant to our experience. A case might be made that in such a situation, the three-dimensionality of the world is an illusion, but because there is no direct evidence for string theory, I set this possibility aside as speculation. Finally, quantum mechanics represents the state of the world via a $3N$-dimensional wave function, where N is the number of particles in the universe. Quantum mechanics is not just speculation; it is a well-confirmed theory. Here perhaps there really is evidence that the three-dimensionality of the world is illusory. To see whether this surprising conclusion is warranted, we first need to understand why quantum mechanics requires a $3N$-dimensional wave function.

1 The Wave Function

Consider a very simple world, consisting of just two particles moving around in three-dimensional space. We can pick an origin and three mutually orthogonal directions, and hence impose a coordinate system on the space; the state of the world at a time can then be represented as two points in this space, say, (1, 4, 3) and (9, 2, 6). Alternatively, we can represent the same state as a single point in a six-dimensional space, namely, (1, 4, 3, 9, 2, 6). The latter is called a *configuration space* representation, because each point in the space represents a configuration of particles in ordinary three-dimensional space. Pretty clearly, the two representations are equivalent—any arrangement of particles in three-dimensional space corresponds to a point in configuration space, and conversely any point in configuration space corresponds to an arrangement of particles in three-dimensional space.

Now suppose that instead of particles, our toy world contains two *fields*—continuous distributions of some quantity over three-dimensional space, rather than discrete particles. That is, the three-dimensional space contains two objects—a function $f_1(x, y, z)$ and a second function $f_2(x, y, z)$, each representing the intensity of its respective field as a function of the three spatial coordinates. Again, we can represent the same state in a six-dimensional space as a function $F(x_1, y_1, z_1, x_2, y_2, z_2)$, where the first three coordinates are those of the field f_1 and the second three are those of f_2. For example, suppose field f_1 has nonzero intensity only in two regions A and B, and f_2 has nonzero intensity only in two regions

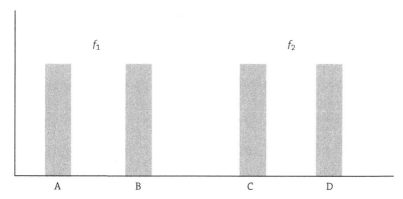

Figure 5.1 Field intensity in 3D space

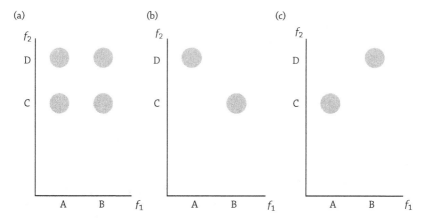

Figure 5.2 Field intensity in 6D space

C and D, as shown (schematically) in figure 5.1. Here the horizontal axis represents the three spatial dimensions, reduced to one for ease of representation.

Alternatively, we can choose to represent the state of the system in the six-dimensional space shown (schematically) in figure 5.2a. Here the horizontal axis represents the coordinates of f_1, the vertical axis represents the coordinates of f_2, and the shaded areas represent nonzero field intensity. In the f_1 coordinates, the field has nonzero intensity only in regions A and B, and in the f_2 coordinates, the field has nonzero intensity only in regions C and D, as in the three-dimensional representation. But note that exactly the same is true of the field distributions shown in figure 5.2b and 5.2c. That is, any one of these six-dimensional field distributions (and many others besides) can adequately represent the three-dimensional field distribution of figure 5.1.

So any two fields in three-dimensional space can be represented by a field in six-dimensional space—and in fact there is considerable freedom in choosing the latter, because many such representations contain all the information in the three-dimensional representation. By the same token, a field in a six-dimensional space in general cannot be represented without loss as two fields in three-dimensional space, because the six-dimensional field contains information that is not present in the corresponding pair of three-dimensional fields. This is reflected in the fact that the distinct six-dimensional fields of figure 5.2a, 5.2b, and 5.2c correspond to one and the same three-dimensional representation—that of figure 5.1.

In quantum mechanics, the state of a system of N particles can be expressed as a function of $3N$ coordinates, three for each particle. This is the wave function, written $\psi(x_1, y_1, z_1, \ldots x_N, y_N, z_N)$. As this expression suggests, the wave function is most naturally represented in a $3N$-dimensional space; in fact, the wave function cannot be expressed as a set of N fields in three-dimensional space because the empirical predictions of the theory depend on the information that is lost in the three-dimensional representation. For example, for a two-particle system, the three wave functions depicted in figure 5.2 would result in different predictions for the measured positions of the particles; figure 5.2c represents a state in which particle 2 is at C if and only if particle 1 is at A, figure 5.2b represents a state in which particle 2 is at C if and only if particle 1 is at B, and figure 5.2a represents a state in which there is no such correlation between the locations of the particles.

The foundations of quantum mechanics are notoriously contested, and the status of the wave function is no exception. But there is a prima facie case, at least, that the wave function should be regarded as the fundamental entity of quantum mechanics. The evolution of the wave function is governed by the basic dynamical law of the theory, and the wave function at the end of an experiment generates its empirical predictions. As J. S. Bell famously put it, "No one can understand this theory until he is willing to think of ψ as a real objective field ... Even though it propagates not in 3-space but in $3N$-space" (1987, p. 128). That is, realism in the context of quantum mechanics arguably commits you to the existence of an entity corresponding to the wave function of the world—a $3N$-dimensional field, where N is the total number of particles in the universe. This is a radical violation of the intuition that there are three spatial dimensions—it is not that there are 4, or 10, but that there are at least 10^{80} spatial dimensions.

2 Keeping up Appearances

Why does it look to us as if there are three spatial dimensions, if in fact there are not? David Albert (1996) suggests that the answer lies in the dynamical law by which the quantum state evolves; an element of the dynamical law called the Hamiltonian takes a particularly simple form if the $3N$ spatial coordinates are grouped into N sets of three (rather than $3N/2$ sets of two, or $3N/4$ sets of four, etc.). The Hamiltonian is a representation of the energy properties of a physical system. In classical physics it is a function from the positions and momenta of the particles in the system to a real number representing the total energy of the system—kinetic energy plus potential energy. The use of the Hamiltonian in classical mechanics is somewhat arcane—it is not the high school version of the theory—but it allows for a particularly succinct expression of the dynamical laws by which the state of the system evolves over time. In quantum mechanics, on the other hand, the Hamiltonian formulation of the theory is the canonical one. Properties are represented as operators rather than functions in quantum mechanics; the Hamiltonian is a differential operator acting on the wave function. But as in classical mechanics, the Hamiltonian acting on a wave function represents the energy properties of the system with that wave function, and as in classical mechanics, this representation allows for a very succinct expression of the dynamical law by which the wave function evolves. This dynamical law—the Schrödinger equation—says that (up to a constant) the rate of change of the wave function is equal to the result of applying the Hamiltonian operator to the wave function. In brief, for both classical and quantum mechanics, the Hamiltonian tells you how the total energy varies across possible states of a system and plays a key role in one formulation of the dynamics for the system.

With this understanding in mind, let us return to Albert's argument. Note first that there is a sense in which classical behavior emerges from quantum mechanical behavior in the macroscopic limit—that is, as systems become large and complicated. The sense is that while microscopic systems must typically be represented by a spread-out wave function in configuration space, macroscopic systems can always be represented to a good degree of approximation by a *point* in configuration space. A point in $3N$-dimensional configuration space, as already shown, can be represented equally well as N points in three-dimensional space. Of course, it can also be represented as $3N/2$ points in a two-dimensional space, and so on. But if we choose to group the coordinates into threes, then the Hamiltonian takes a particularly neat form; the potential energy term depends only on the distances between the N "particles." On the other hand, if we choose to group the coordinates into twos or fours or sevens, then the potential energy term will bear no straightforward relation to the "interparticle" distances so produced.

The key point is that the form of the Hamiltonian one obtains by grouping into threes corresponds to our classical world-view; potential energy, be it gravitational, electrostatic, or some other, is taken to be generated by forces that depend only on the distances between the objects involved. Indeed, elementary classical mechanics takes these interparticle force laws (e.g., the inverse-square laws of gravitation and electrostatics) to be fundamental, and the Hamiltonian (if one thinks of it at all) is taken to be a construction based on these fundamental laws. If Albert is right, we have to give up this picture, because there are no particles moving in three-dimensional space and hence no laws governing their motion; it is the wave function and Hamiltonian in $3N$ dimensions that are fundamental. But his claim is that we can nevertheless explain why our intuitive picture of the world has three dimensions, because it is the obvious (though false) interpretation of the behavior of macroscopic objects. That is, even though there is nothing in fundamental reality corresponding to our choice of coordinate grouping, if we choose to group the coordinates in threes, a particular description of the behavior of medium-sized everyday objects becomes available to us, namely, the *classical* description of objects moving in three-dimensional space, subject to forces that depend on the distances between them. Hence "quantum-mechanical worlds are going to appear (falsely!) to their inhabitants, if they don't look too closely, to have the same number of *spatial dimensions* as their *classical counterparts* do," namely, three (Albert 1996, p. 282). The reason for the caveat, of course, is that if we *do* look closely—if we perform experiments that reveal the underlying quantum mechanical nature of microscopic reality—then we convince ourselves that the world cannot really be three-dimensional via the arguments of the previous section.

If Albert is right, then the world is really $3N$-dimensional yet appears three-dimensional to us. But is he right? There are reasons to be skeptical (Monton 2002; Lewis 2004). First, note that the story about grouping the coordinates into threes is somewhat simplified; more is required to recover the coordinate system in which the Hamiltonian takes the familiar three-dimensional form. Consider a classical two-particle system, represented by a point in a six-dimensional space. The choice of origin and the order in which we list the coordinates is conventional; we can choose any system we like. But starting from such a coordinate system, we cannot recover the three-dimensional coordinates in which the standard force laws hold simply by grouping the coordinates into threes. First, we have to make sure that the axes line up—that the two x-axes, for example, point in the "same" direction. Second, we have to make sure that the coordinates are ordered in the "right" way, such that the first and fourth are the x-coordinates of the (fictional) particles, the second and fifth are the y-coordinates, and the third and sixth are the z-coordinates. Finally, we have to make sure that the origins for the two x-coordinates "coincide," and similarly for the y- and z-coordinates. Note all the scare quotes here. The two x-axes cannot really point in the same direction, because they are orthogonal in configuration space; nevertheless, one particular relative orientation is required for the Hamiltonian to take the right form. Similarly for the ordering and the origin; no coordinates are really x-coordinates, so the origins of the two x-coordinates cannot really coincide. If there isn't really a three-dimensional space, it seems that there is no explanation for these requirements other than that they work to produce the familiar Hamiltonian. Indeed, one might start to suspect that requirements like this are part of what it takes for a space to be three-dimensional.

On a related note, our picture of force laws in three dimensions is more than just an expendable intuition; it is the explanation of the form of the Hamiltonian.

It is not just high school physics that takes force laws between three-dimensional objects as fundamental; when quantum mechanics is applied to a physical system, the potential energy term in the Hamiltonian is constructed by appeal to such laws. Under Albert's way of looking at things, we would have to forgo this explanation of what we are doing. Albert is free to postulate that the Hamiltonian is unexplained and fundamental, but then it becomes mysterious why the method of constructing the Hamiltonian from the three-dimensional configuration of our system works at all.

Finally, although grouping the coordinates into threes yields a particularly simple Hamiltonian, this simplicity cannot be parlayed into an explanation of three-dimensional appearances. It is true that a particular choice of coordinates sometimes plays a role in the explanation of appearances. For example, I might choose coordinates in which my office is at rest (rather than ones in which the sun is at rest) to explain my experience of thunderclouds building behind the skyscrapers of downtown Miami. But the dependence of the explanation on the choice of coordinates here is superficial; it is simply a matter of calculational convenience. Even if I used the sun-centered coordinates, it would still be the case that the clouds, the buildings, and my office lie in a straight line. The coordinate dependence in Albert's account is much more thoroughgoing—to an extent that makes it problematic. For one thing, three points that lie on a straight line under one grouping of $3N$-dimensional coordinates will not, in general, lie in a straight line under another grouping. For another, the patterns of points that we identify as buildings and clouds will not, in general, be present under another grouping. But patterns that appear only under one arbitrary choice of coordinates are generally regarded as artifacts of that choice rather than facts about the world, because the facts about the world presumably do not depend on our representational choices. If there are no buildings, clouds, and offices that lie in a straight line, then the explanation of my experience evaporates.

Now we seem to have painted ourselves into a corner. The state of a quantum mechanical system cannot be represented without loss in three dimensions; it has to be represented in a $3N$-dimensional configuration space. But the evolution of the wave function in $3N$ dimensions cannot explain our three-dimensional appearances. So either quantum mechanics is wrong, or we are radically deceived about the nature of our own experience. Neither of these looks like an attractive option.

3 Configuration Space

All is not lost, because there is another option available to us—namely, that we are wrong about the structure of space. A rather flatfooted solution along these lines would be to propose that there are in fact two spaces—the $3N$-dimensional space in which the wave function evolves, and the three-dimensional space in which the objects of our experience move around—neither of which can be reduced to the other. But this proposal introduces as many problems as it solves. How does the motion of the wave function in one space give rise to the motion of objects in a completely separate space (Monton 2006)? A more subtle solution is that the three-dimensional structure is already present in the configuration space. Thus far, we have been treating the space in which the wave function evolves as $3N$-dimensional in the same sense that ordinary space is three-dimensional—essentially, as a space spanned by $3N$ mutually orthogonal vectors and having no other structure. Perhaps there's more to configuration space than initially meets the eye.

Consider ordinary three-dimensional space. To impose coordinates on this space, you choose an origin, a length scale, and three mutually orthogonal axes. We have been supposing that the coordinatization of configuration space is just like this; to impose coordinates on a $3N$-dimensional configuration space, you choose an origin, a length scale, and $3N$ mutually orthogonal axes. This freedom in choosing coordinates leads to the problems facing Albert's argument, because most such choices do not yield the simple Hamiltonian or the objects of ordinary experience.

Suppose instead that we take seriously the idea of a configuration space as a space of configurations—that is, a space that is intrinsically structured as N sets of three-dimensional coordinates. Mathematically, this is not hard to do. Instead of modeling the space as an ordered $3N$-tuple of parameters, $\langle x_1, x_2, \ldots, x_{3N} \rangle$, we model it as an ordered N-tuple of ordered triples:

$$\langle \langle x_1, y_1, z_1 \rangle, \langle x_2, y_2, z_2 \rangle, \ldots, \langle x_N, y_N, z_N \rangle \rangle.$$

Rather than specifying the coordinates by choosing $3N$ axes, we choose three—the x-, y-, and z-axis, which are the same for each triple. That is, x_1 through x_N pick out points on the same axis, and similarly for y and z. Then the wave function can be regarded as a function of these parameters—as a mathematical entity inhabiting a $(3 \times N)$-dimensional configuration space, rather than a $3N$-dimensional plain space. And the basic thesis of wave function realism is that the world has this structure—the structure of a function on $(3 \times N)$-dimensional configuration space.

Given that configuration space has this structure, an Albert-style appeal to dynamics to generate three-dimensional appearance is impossible, but it is also unnecessary. It is impossible because the Hamiltonian takes exactly the same form under every choice of coordinates, so no choice makes it especially simple. It is unnecessary because the outcome of that argument—that the coordinates are naturally grouped into threes—is built into the structure of reality, and hence doesn't need to be generated as a mere appearance based on the simplicity of the dynamics.

4 Dimension without Illusion

The proposal of the previous section is designed to allow us to say that the world of quantum mechanics really is three-dimensional, and hence the three-dimensionality of appearances does not have to be generated as any kind of illusion. Our appearances are veridical. But does the proposal really allow us to do that? The wave function is still a function of 3*N* independent parameters, even if those parameters have some internal structure. Isn't that a prima facie reason to say that the wave function—and hence the world—is 3*N*-dimensional?

Previously, I took the lesson here to be that the term *dimension* is ambiguous in the quantum mechanical world (Lewis 2004). One can take it to refer to the number of independent parameters required to specify a point in the space in which the quantum state evolves, or one can take it to refer to the number of independent axes required to impose coordinates on the space. In the classical case, these two coincide; a point in the space in which the classical state evolves is specified by three parameters, and imposing coordinates on the space requires three axes. But in the quantum case, the two dimension concepts come apart; it takes 3*N* parameters to specify a point in the space in which the wave function evolves, but only three axes are required to impose coordinates on this space. Hence, if quantum mechanics is true, there is a sense in which the world is 3*N*-dimensional and a sense in which it is three-dimensional. Under the latter sense, the apparent three-dimensionality of the world is no illusion.

My 2004 position is essentially the same as the one Albert adopts in the coda to his 1996 paper (although our reasons for adopting the position are different). Albert restates his thesis "a bit more diplomatically" (1996, p. 282). The diplomatic version of his thesis is that there are "two ideas we're accustomed to having in mind when we think of 'physical space.'" The first is "the space of possible interactive distances"—that is, the space in which the distances between points are the "interparticle distances" appearing in the Hamiltonian. This space is three-dimensional—the dynamical laws are laws of a three-dimensional world. The second is the space "in which a specification of the local conditions at every address at some particular time (but not at any proper subset of them) amounts to a complete specification of the physical situation of the world, on that theory, at that time." This space is 3*N*-dimensional; it takes a specification of the wave function amplitude at every point in a 3*N*-dimensional space to completely specify the quantum state of the world. So, diplomatically speaking, it is not that the three-dimensionality of the world is an illusion; rather, the world really is three-dimensional under one reading of "dimension," and it really is 3*N*-dimensional under the other reading.

This position still seems tenable to me, but there is an alternative that I now find more attractive. Albert initially claimed that the world is 3*N*-dimensional, and the appearance that the world is three-dimensional (when we don't look too closely) is illusory. Later he claimed (and I concurred) that the three-dimensional appearances (when we don't look closely) and the 3N-dimensional appearances (when we do) can both be taken as veridical, because "dimension" is ambiguous. However, there is a third possibility—namely, that the world is three-dimensional, and the 3N-dimensional appearance of the world when we look closely is illusory. That is, our everyday impression that the world is three-dimensional is correct, but the impression we get from our quantum mechanical experiments that the world is 3N-dimensional is misleading. This third position seems to have something to be said for it, and in what follows I explore and defend it.

5 Dimensions and Parameters

The claim that the quantum world is 3N-dimensional is based on the fact that the wave function is a function of 3N independent parameters. This is a perfectly good characterization of dimension in some general sense of the concept. But the number of independent parameters is not always a good way to characterize the spatial dimensionality of a system. Sometimes a parameter is not a spatial parameter at all. For example, in evolutionary game theory, the state of a population of organisms can be represented as a function of n parameters, one for each organism, in which each parameter represents the continuum of possible strategies the organism can adopt in interacting with the others, and the function represents the probability distribution over the strategy space—that is, the chance that organisms will use particular strategies in their interactions with each other. Clearly nothing should be inferred from this model about the (literal, rather than figurative) space in which the organisms live.

Of course, in the quantum case the parameters are spatial; each is a position coordinate for a particle. Still, it does not follow that the representation is of a spatially 3N-dimensional world. Consider again the case of evolutionary game theory. Insofar as such a model is intended to be realistic, it is intended to be about n individual organisms, each with its own probability distribution for adopting a particular strategy. But there may be information in the full n-dimensional representation that is lost when the population is represented as n single-organism states. We can reinterpret figures 5.1 and 5.2 to demonstrate this. Suppose figure 5.1 represents the strategy distributions for two organisms; organism 1 adopts strategies of type A and type B with equal probabilities, and organism 2 adopts strategies C and D with equal probabilities. But the single-organism properties of figure 5.1 are compatible with each of the three n-organism states represented in figure 5.2. Figure 5.2a represents a state in which the strategy adopted by each organism is independent of that adopted by the other organism, and figure 5.2b and 5.2c represent states in which the strategy adopted by each organism depends on what the other does. Clearly the future evolution of the system depends on which n-organism state the population has. But this wouldn't warrant the inference that reality consists of an n-organism entity rather than n

individual organisms. The organisms just happen to have complicated conditional properties; their strategies depend on those of the organisms with which they interact. The most convenient way to represent these conditional properties— perhaps the only convenient way—is via the n-dimensional state, but nothing ontological should be read into this.

One can adopt a similar position with regard to the quantum mechanical wave function. That is, the fact that the state of a quantum system can be represented without loss as a single wave function in a $3N$-dimensional configuration space but not as N single-particle wave functions in three-dimensional space does not entail that the world is spatially $3N$-dimensional. Rather, one could interpret the situation as one in which the particles have complicated conditional connections; the position properties of one particle, as encoded in the wave function, depend on those of the other particles (see Monton's chapter in this volume). The $3N$-dimensional wave function may be the only convenient way to represent these properties, but it does not follow that it literally represents the spatial structure of the world. This is not to give up the assumption of wave function realism; the structure of the wave function accurately reflects the structure of the world, but some of that structure consists in correlations between spatial positions. The wave function is a function of $3N$ parameters, but we need not interpret each parameter as an independent spatial direction.

6   Spatial Phenomena

So far, I have argued that we *need* not interpret the quantum world as spatially $3N$-dimensional; we cannot simply read the spatial structure of the world off the mathematical structure of our representation. But this leaves open the question of how we *should* interpret the wave function. The question is whether the $3N$ parameters of configuration space deserve to be called spatial dimensions. The answer, I suggest, hangs on the connections between those parameters and spatial phenomena.

The claim that the quantum world is three-dimensional is based on a fairly direct correspondence between the structure of configuration space and the structure of our spatial experience. I can stick my arm out in some arbitrary direction and stipulate that every third configuration space parameter refers to that direction, and similarly pick two directions orthogonal to my arm to correspond to the remaining configuration space parameters. The sense in which the world is three-dimensional is straightforwardly a spatial sense. But no such direct correspondence to experience is available for the claim that the quantum world is $3N$-dimensional; if the first and fourth configuration space parameters are in fact orthogonal spatial directions in some sense, this is no part of my experience. Hence the doubt that the $3N$ parameters should be characterized as spatial.

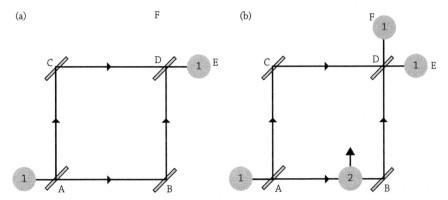

Figure 5.3 Mach-Zehnder interferometer

Still, an indirect correspondence to experience might be available. Consider again the Flatland scenario that we started with. The two-dimensional Flatlander suspects that he is really living in a three-dimensional world because various phenomena he observes are explicable in terms of three dimensions, but not in terms of two. Similarly, various phenomena in the quantum world are explicable on the assumption that the quantum state evolves in a $3N$-dimensional space, but arguably not under the assumption that it evolves in a three-dimensional space. Consider, for example, the Mach-Zehnder interferometer shown in figure 5.3a. A single-particle wave packet (labeled 1 in the diagram) enters the device at the bottom left and is split into two equal wave packets by the half-silvered mirror A. The two packets are reflected by mirrors at B and C, respectively, and both arrive at the second half-silvered mirror D. If the path lengths ABD and ACD are exactly equal, then the wave components emerging toward detector F are exactly out of phase—the troughs of one wave coincide with the peaks of the other—and the two waves exactly cancel out. By the same token, the wave components emerging toward detector E are exactly in phase—the peaks coincide and the troughs coincide—and the two waves add in intensity. The result is a single wave packet of exactly the intensity of the one input at the bottom left, emerging toward detector E. That is, the two wave packets traveling via B and C collide and interact at D, producing a single wave packet traveling toward E.

Now suppose that the wave packet traveling via B interacts with a wave packet associated with another particle (labeled 2 in the diagram), and the wave packet traveling via C does not, as shown in figure 5.3b. What happens now is that the two wave packets arriving at D no longer interact, resulting in wave packets traveling onward toward E and F. It is as if the wave packets pass by each other rather than colliding. Indeed, if one adopts a configuration space representation, this is exactly what happens; the packet traveling via B undergoes a shift in the coordinates of the second particle, whereas the packet traveling via C undergoes no such shift. I have attempted a configuration space representation in figure 5.4.

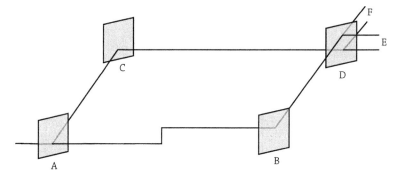

Figure 5.4 Configuration space representation

The full configuration space has six dimensions; I have chosen three to include in the diagram, the two horizontal dimensions representing the plane of the experiment in the coordinates of particle 1, and the vertical dimension representing one of the coordinates of particle 2. A wave packet representing both particles enters at the bottom left and is split in the coordinates of particle 1 by mirror A. The vertical jog on the lower path represents the collision of the particles; particle 2 moves, so the trajectory shifts in the coordinates of particle 2. But now the two wave packets do not hit the same spot in the (five-dimensional) plane of mirror D because they differ in the coordinates of particle 2. Hence, no interference occurs at D, and each packet is again equally split into two further packets in the coordinates of particle 1.

Here, then, we apparently have indirect evidence that the quantum world is $3N$-dimensional. The extra dimensions seem necessary to explain the interactions (or lack of them) between wave packets. But this evidence needs to be treated with considerable caution. Note in particular that wave packets corresponding to different particles cannot pass by each other in the way just described; if two wave packets corresponding to two different particles approach the same region of three-dimensional space, they interact. Consider the interaction between two particles depicted in figure 5.5. Figure 5.5a is the ordinary three-dimensional representation; the wave packets approach and bounce of each other, and this is the case even if particle 1 or 2 interacts with some other particle on its way to the collision point. Figure 5.5b depicts the same interaction in configuration space, with the vertical axis representing the x-coordinate of particle 1 and the horizontal axis the x-coordinate of particle 2. The two-particle wave packet travels up to the diagonal line where the coordinates of the two particles coincide, and here the trajectory suddenly changes direction. The trajectory in configuration space acts as if it bounces off a reflective wall, a wall that extends indefinitely in the coordinates of any other particles involved. Of course there is no physical wall; this behavior is purely an expression of the dynamical laws. Pictured in configuration space, these laws look mysterious; why should the trajectory sud-

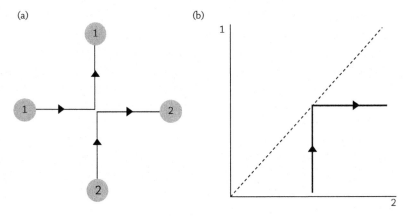

Figure 5.5 Interaction between two particles

denly change direction just there? Pictured in three-dimensional space, the reason for the sudden turn is obvious—the particles collide. The phenomena are not what one would expect of objects moving in a 3N-dimensional space, but they are precisely what one would expect of objects moving in a three-dimensional space.

Of course, Albert might reply that I have switched back to the sense in which configuration space is three-dimensional—namely, that the dynamical laws are those of a three-dimensional space. Quite right, but my contention here is precisely that the usual sense of the term *spatial* is intimately connected with dynamical laws. Spatial phenomena concern whether or not objects bounce off one another or pass by one another—and these are dynamical concepts. When we are dealing with wave packets for distinct particles, three-dimensional space, not 3N-dimensional space, is the arena in which spatial phenomena play out. What's more, the interactions of ordinary macroscopic objects play out in three-dimensional space, and this kind of behavior ultimately underlies our ordinary use of the term *spatial*.

Admittedly, consideration of the example with which we started this section complicates this story. Two wave packets that are components of the state of one and the same particle sometimes interact and sometimes pass by each other when their three-dimensional coordinates coincide. Doesn't this require the existence of extra dimensions in which the passing by can take place? The lesson of the previous section is that such phenomena do not require the existence of extra spatial dimensions. Certainly one needs parameters in the theory, the values of which determine whether the packets interact. In the quantum case, the parameters in question refer to the coordinates of the other particles in the system—that is, they encode how the wave packet for the particle we are following is correlated with the wave packets for the other particles. The question is whether we should regard this structure of parameters as spatial. We can now see clearly the costs and benefits of each answer. If we think of spatial phenomena as 3N-dimensional, we get a nice literal interpretation of the operation of the interferometer (figure 5.4), but the behavior of everyday objects becomes mysterious (figure 5.5b). If we think of spatial phenomena as three-dimensional, then we can go on thinking of everyday phenomena in terms of collisions between objects (figure 5.5a), but the operation of the interferometer has to be taken as involving correlation rather than spatial location, because the spatial locations don't determine the particle's behavior (figure 5.3). To my eye, the choice seems clear; the former choice does considerable violence to our ordinary spatial notions, whereas the latter just requires a slightly less intuitive understanding of interference phenomena. Even in the latter case, it's not as if we cannot use the 3N-dimensional representation to picture interference to ourselves; we just can't take that picture as a direct *spatial* representation.

7 The Pragmatic Dimension

The world looks three-dimensional unless one looks closely, when it looks 3N-dimensional. But which appearance is veridical and which the illusion? Albert initially contends that the three-dimensionality of the everyday world is illusory, and that the 3N-dimensional wave function one discerns in quantum phenomena is the reality behind the illusion. What I have tried to do here is to argue for the converse of Albert's initial position; the world really is three-dimensional, and the 3N-dimensional appearance of quantum phenomena is the theoretical analog of an illusion; we represent quantum reality to ourselves as 3N-dimensional to more readily visualize the correlations between wave packets. The basic reason for thinking of things this way is that the sense in which the quantum world is three-dimensional corresponds directly to the way we already use the term *spatial*, whereas the sense in which it is 3N-dimensional does not.

Still, even if it would be a stretch to apply the adjective *spatial* to the 3N dimensions of configuration space, there is nothing to stop us from doing so; it would not be the first time that ordinary usage has shifted as a result of scientific advances. This is ultimately a pragmatic matter. But I find it hard to see any motivation for such a shift, because the phenomena in question are so far removed from everyday life. We modified our concepts of rest and motion to accommodate the idea that Earth is moving, but doing so also gave us tools to better describe everyday phenomena. If a shift in our concept of spatial dimension would reap similar benefits, then no doubt such a shift should and will occur. Until then, we can say with a straight face that the world is three-dimensional.

References

Abbott, Edwin A. (1952), *Flatland*. New York: Dover.

Albert, David Z. (1996), "Elementary Quantum Metaphysics," in J. Cushing, A. Fine, and S. Goldstein (eds.), *Bohmian Mechanics and Quantum Theory: An Appraisal*. Dordrecht: Kluwer, pp. 277–84.

Bell, John S. (1987), *Speakable and Unspeakable in Quantum Mechanics*. Cambridge: Cambridge University Press.

Lewis, Peter J. (2004), "Life in Configuration Space," *British Journal for the Philosophy of Science* 55:713–29.

Monton, Bradley (2002), "Wavefunction Ontology," *Synthese* 130:265–77.

Monton, Bradley (2006), "Quantum Mechanics and 3N-Dimensional Space," *Philosophy of Science* 73:778–89.

6

The Nature of the Quantum State*

TIM MAUDLIN

For Plato and Aristotle, the fundamental philosophical question about anything is *ti esti*: what is it? The same question has recurred throughout the history of physics. Somewhat surprisingly, physics has at times managed to develop powerful and predictively accurate theories even in the absence of any answer to this fundamental question. A few examples may help focus our attention on the how physics can proceed without any understanding of the nature of the relevant physical entities and also on the various forms that an answer to this fundamental question might take.

During the early development of thermodynamics, Fourier provided an exact and mathematically detailed account of how heat behaves without providing any clear account of what heat is.[1] Common experience confirms that heat always flows from a hotter to a colder body, and by detailed observation Fourier arrived at exact differential equations for heat flow, but the dynamics of heat does little to shed light on its nature. The caloric theory took "flows" literally, postulating that heat is a kind of fluid. According to this theory, there would be nothing to prevent the existence of a container full of heat and nothing else. According to this theory, heat is a kind of substance.

This basic ontological categorization of heat naturally gave rise to other questions. If heat is a fluid, is cold another kind of fluid (frigoric), or is cold just the absence of caloric? Although the question sounds odd to modern ears, it is no more absurd than the debate among early electricians between the "one fluid" and "two fluid" theories, a debate that the two-fluid side properly won.

Opposed to this, of course, is the kinetic theory of heat. If heat is just the microscopic motion of atoms and molecules, then there cannot be a container full of just heat, and it is hard to see how cold could be anything but a low degree

---

* Thanks to Shelly Goldstein for comments and corrections.

[1] Indeed, the one thing Fourier positively declared about the nature of heat was wrong: "But whatever may be the range of mechanical theories, they do not apply to the effects of heat. These make up a special order of phenomena, which cannot be explained by the principles of motion and equilibrium" (Fourier 1952, p. 169).

of microscopic motion. So the answer to a basic ontological question can direct further research in a straightforward way.

In this example, the two competing theories located heat in preexisting ontological categories: substance and motion. But sometimes the progress of science expands the set of conceptual possibilities. Electromagnetism provides a case in point. First one had to decide whether to take the regions surrounding charged or magnetized bodies as containing anything physical at all. Faraday (1952) argued in "On the Physical Lines of Magnetic Force" that the aforementioned lines should be considered as *physical* rather than merely *mathematical*. He contrasted the gravitational force that the sun exerts on Earth with its "illuminating or warming power." Each of these could be associated with certain lines of force, in particular with straight lines that originate at the sun and terminate at Earth. But the gravitational force of the sun on Earth is unaffected by material in the intervening region, whereas the lines of the force of illumination (as we would say, the light rays) can be altered in direction, refracted, reflected, and so on. Similar observations show that magnetic "lines of force" have a physical, rather than just a mathematical, nature.

That still leaves open a lot of distinct possibilities. The electrical and magnetic lines of force might, for example, be stresses and strains in a material medium (the electromagnetic ether), or they might be trajectories of some particles, or regions where chains of dipolar molecules have lined up, or something else altogether. Just as Fourier's detailed equations for heat flow cannot, of themselves, settle the question of the nature of heat, so detailed equations that describe the behavior of the electromagnetic field may not offer any obvious purchase on its fundamental nature. Rather, each proposal leads to a distinct set of further questions and experiments. If the lines of force are trajectories of particles, what are the masses and velocities of the particles? If dipolar molecules are lining up (so the question of velocity does not arise), what is the strength of polarity, and the density of the molecules? Most famously, if there is an ether, what are its mechanical properties? Is it entrained by Earth, or does Earth pass through it, creating an ether wind? Can the ether be rarified or compressed, or even evacuated from a spatial region?

The example of the electromagnetic lines of force has an instructive conclusion. The possible explications offered here were all drawn from preexisting physical categories. Yet in the end, the question of the nature of the lines of force was settled (at least temporarily) by inventing a whole new species of physical entity: the field. A field is not a collection of particles, does not depend on the presence of polarized molecules, and does not require a mechanical ether. But what, we might hear a nineteenth-century philosopher or physicist ask, *is* it? What category of being does it belong in?

By now, we have become so accustomed to the notion of a field that we are likely to be frustrated and answer roughly. There is no necessity that an electromagnetic field should fit neatly into any preexisting ontological scheme: there are more things in heaven and Earth, Horatio, than are dreamt of in your philosophy. One should stop trying to assimilate the field to some familiar concept, and rather admit it as a fundamental new sort of physical entity.

I think this is an eminently reasonable first step toward understanding the electromagnetic field, and hence toward understanding the status of Faraday's lines of force. But it is only a first step. The new ontological category "field" has to be more than just a space in a categorical scheme with a name attached. The question is how we come to fill out the new concept with more positive content. What are our resources for coming to a more detailed account of what a field is?

In the case of the electromagnetic field, we have several, disparate resources. In the first place, we start with a *mathematical representation* of the field. The mathematical representation could take many forms: a pair of vector fields, a single tensor, a connection on a fiber bundle, or even, as Faraday had, a collection of lines in space. However, we must guard against the temptation to simply reify this mathematical representation and accept without proper investigation the idea that the mathematical features of the representation directly correspond to physical features of some real entity.

Faraday was concerned in the first place with perhaps the most basic such feature: location in space. In the course of solving a problem in gravitational mechanics, Newton would make use of a line drawn between two gravitating bodies, the line that specifies the direction of the gravitational force. *The line*, as a mathematical object, occupies a continuous region connecting the two bodies. Should we therefore assume that there is anything physical in that region? After noting that the gravitational force is unaffected by other circumstances (other bodies may exert a gravitational force on these, but those additional forces do not change or disrupt the force attributed to these bodies alone), Faraday concludes:

> In the case of gravitation, no effect which sustains the idea of an independent or physical line of force is presented to us; as far as we at present know, the line of gravitation is merely an ideal line representing the direction in which the power is exerted. (Faraday 1952, p. 817)

My aim is not to defend Faraday's particular conclusions. He was properly circumspect about his ideas, and in some cases the basis for his judgment turned out to be incorrect. For example, he regarded the gravitational influence as instantaneous: "We have no evidence that *time* enters in any way into the exercise of this power, whatever the distance between the acting bodies, as that from the sun to the earth, or from star to star" (Faraday 1952, p. 819).

If Faraday had been convinced that the same eight minutes required for light to get from the sun to Earth is also required for the gravitational effects of events on the sun to influence Earth, he might have reconsidered his opinion about the lines of gravitational force. My point is merely that he recognized the gap between use of a mathematical object in a theory and the existence of a physical entity that somehow corresponds to or is represented by that object. He put some effort into considering the sorts of circumstances that suggest the existence of a physical entity rather than just a convenient mathematical instrument.

When we turn to quantum theory, this need to separate the mathematical representation from the physical entity represented becomes particularly pressing. The use of sophisticated mathematical constructs in physics has become so pervasive and intimidating that the very distinction between the representation and the thing represented threatens to disappear from view. Just as Faraday needed a distinction between physical lines and ideal (merely mathematical) lines to raise his ontological question, we need to keep the distinction between mathematical and physical entities sharp. Unfortunately, the usual terminology makes this difficult.

The topic we are trying to address is, more often than not, referred to as "the ontological status of the wave function." But one obvious thing about the wave function is that it is a *function*, that is, a mathematical object. In the mathematics of standard nonrelativistic quantum mechanics, there is such a function, whose domain is something called configuration space and whose range is the complex numbers. When we express the dynamics of this theory using, for example, Schrödinger's equation, that equation applies to such a *mathematical* object. In this sense, the question of the ontological status of the wave function is trivial: it is a piece of mathematics. (We can, and should, worry about the exact status of configuration space, but no matter how that comes out, the wave function is still a mathematical map from it to the complex numbers.)

Of course, that is not at all the question we have in mind. Rather, the question is what, if anything, does this particular piece of mathematics *represent*, what physical entity (if any) corresponds to it, and how is that correspondence to be understood? This second item, the entity *represented* by the mathematical wave function, has no standard name. Most often, it is simply also called the wave function of a system. Although it may be possible to tell by context exactly which of these two items is meant, the ambiguity makes our topic unnecessarily confusing. To forestall such confusions, I have opted for the term *quantum state* to refer to the physical entity. This is the thing we want to understand.

Since the wave function is the mathematical representation of the quantum state, one of our central questions concerns the nature of the representation. The simplest possibility is that the collection of mathematical items and the collection of possible physical states stand in a one-to-one relation: each mathematical item represents a single physical possibility, and each physical possibility is represented by exactly one mathematical object. If this obtained, then it would be trivial to understand how a mathematical equation such as Schrödinger's equation could represent the dynamics of a real physical thing. Solutions to Schrödinger's equation are continuous sequences of wave functions indexed by a parameter t. Taking t to represent nonrelativistic time (given an arbitrary choice of time unit and origin), such a mathematical function from t to the space of wave functions would represent the change of the quantum state through time. Study of these mathematical solutions would then at least indicate what the quantum state is *doing*—much as Fourier's equations describe the flow of heat—even if it does not directly reveal what the quantum state *is*.

However, matters are not so simple. The earliest text I know of that claims a one-to-one relation between quantum states and wave functions is David Albert's "Elementary Quantum Metaphysics":

> The sorts of physical objects that wave functions *are*, on this way of thinking, are (plainly) *fields*—which is to say they are the sorts of objects whose states one specifies by specifying the values of some set of numbers at every point of the space where they live, the sorts of objects whose states one specifies (in *this* case) by specifying the value of *two* numbers (one of which is usually referred to as an *amplitude*, and the other as a *phase*) at every point in the universe's so-called *configuration* space.
>
> The values of the amplitude and the phase are thought of (as with all fields) as intrinsic properties of the points in the configuration space with which they are associated. (Albert 1996, p. 278)[2]

If one takes this characterization literally, then there would be an exact one-to-one relationship between the mathematical representations and the possible quantum states: any difference in mathematical representation would mean a difference in the complex numbers assigned to a point in the configuration space, and hence a difference in an intrinsic physical property at the denotation of that point. Leaving aside the status of "configuration space" in this picture (to which we return later), no standard account of quantum mechanics accepts that there are so many physically distinct possible quantum states. For example, if we change the mathematical wave function globally by a constant phase shift, on the view sketched before, the new wave function would represent a distinct physical possibility, but every standard account of quantum mechanics would insist that the new mathematical object represents the same physical state. That is, quantum states, whatever they are, are supposed to correspond to *rays* in Hilbert space, not to vectors.

It is worthwhile reflecting on the basis for this claim. In the standard approach (whose shortcomings we return to soon), the quantum state is connected to observable phenomena, and hence to empirical evidence, via mathematical operators that are associated with "measurements." In this scheme, the wave function is used together with the operators to derive probabilities for the different

---

[2] See also Lewis 2004 and Albert's and North's contributions to this volume.

possible outcomes of measurements. It is a simple mathematical fact that changing the overall phase of the wave function does not alter those probabilities at all. So if the empirically testable consequences of the theory are exhausted by these probabilities, it follows that any *physical* difference that corresponds to a different overall phase must be completely unobservable.

This conclusion is not limited to the standard presentation. In Bohmian mechanics, for example, and in the Ghirardi-Rimini-Weber (GRW) theory it is also the case that a global change in the phase of the wave function results in no change to the empirical predictions of the theory. In these cases, the empirical predictions are determined by predictions for the location of matter in space-time, but again the overall phase of the wave function does not affect these predictions. What this suggests to most commentators is that wave functions differing only in a global phase factor do not represent different physical possibilities, different quantum states. What it suggests is that the relation of wave functions to possible quantum states is not one-to-one, but many-to-one.

I do not take this inference to be analytical: contrary to old logical empiricist slogans, I see nothing at all impossible about two distinct physical situations that are empirically indistinguishable, even in principle. But the in-principle unobservability of the overall phase of the wave function does rightly pose the question of why it should have been taken to represent a real *physical* degree of freedom in the first place. The suspicion one has is that any such in-principle unobservable physical degree of freedom can be eliminated, and a physical theory with a pared-down ontology but the same empirical consequences produced. If the new theory does not suffer from any other defect relative to the original, most would consider it to be superior.

In the case of the overall phase, how should we adjust the old theory to eliminate any physical degree of freedom that corresponds to that overall phase? There is a trivial way to accomplish this, and a more difficult but more profound way. The trivial method makes no adjustment to the *mathematical* apparatus at all: quantum states are still represented by vectors in Hilbert space, for example. But one adds to the mathematical machinery a postulate about how the mathematics *represents* the quantum states: vectors belonging to the same ray are all taken to represent the same quantum state. The mathematics as mathematics is unaltered; only its physical significance changes.

Treating all the wave functions in a ray of Hilbert space as representative of the same physical state does impose constraints on the mathematics. In particular, it must be a feature of the dynamics, presented as a differential equation for the wave function, that every wave function taken from a given ray evolves, after a certain period of time, into a wave function taken from some single other ray. If this were to fail, and different wave functions from the same ray were to evolve into wave functions from different rays, then the physical states could not be taken to correspond to rays. The linearity of the dynamics, expressed in terms of vectors rather than rays, guarantees this property.

So we have a *necessary* condition for regarding some feature of a mathematical representation as merely a gauge degree of freedom, to which no physical feature corresponds. If we could turn this into a *sufficient* condition as well, then pure mathematical analysis could sort out the gauge degrees of freedom from the mathematical structures that represent real physical features. Unfortunately, things are not so simple. It is possible to partition the mathematical state space into equivalence classes such that every member of one equivalence class evolves, via the dynamics, into a member of some other particular equivalence class, yet the members of a single equivalence class represent different physical possibilities. To take the most extreme example, if one takes the entire mathematical space to belong to the same equivalence class, then our necessary condition will certainly be satisfied, but we do not want to conclude that there is only one possible physical state.[3] We will return to the issue of how to distinguish the parts of the mathematics that represent physical features from those that do not.

Suppose that we have, for whatever reason, decided to regard some aspect of the mathematics as merely gauge, so many distinct mathematical objects represent the same physical situation. Then, as mentioned, there is a second way to respond to the situation: change the mathematics. Instead of using vectors in Hilbert space as the mathematical representation of the quantum state, then bundling them together into equivalence classes that correspond to the same physical situation, why not try to formulate the mathematics directly in terms of a different set of objects, such as the equivalence classes themselves?

The result of such a reformation of the mathematics is to abandon the characteristic structure of Hilbert space altogether. In particular, whereas Hilbert space is a complex vector space in which each vector can be multiplied by a complex number and any pair of vectors can be added to yield a new vector, *projective* Hilbert space (the set of rays in Hilbert space) is not a vector space at all. It makes no sense, for example, to ask what the sum of two elements in projective Hilbert space is. It might be worthwhile to work through the details of this observation.

Let's consider the simplest possible case: the spin of a spin-½ particle. The standard Hilbert space used to represent these possible quantum states is isomorphic to the set of 2×1 complex matrices $\begin{bmatrix} a \\ b \end{bmatrix}$ with a and b complex numbers. We arbitrarily choose some orthogonal triple of directions in space to call the x-, y-, and z-directions. (Note: choosing these directions entails not merely picking three axes in space but also choosing one of the directions along each axis as the "positive" direction. This choice of directions generates one sort of gauge freedom in the mathematics because any other orthogonal triple might have been chosen, yielding a different mathematical representation of the same quantum states. But this is not the gauge freedom that interests us here.)

---

[3] For discussion of one misbegotten attempt to lump together different mathematical representations in this way, see Maudlin 2002.

Having chosen the axes, we can specify what it is to build a Stern-Gerlach apparatus "oriented in the z-direction." This requires us to indicate both the geometry and the polarization of the magnets. (Polarization is conveyed by yet another arbitrary decision to call one pole of a magnet "north," but this also is not the gauge freedom we are interested in.) With all this stage-setting in place, we can define what we mean by, for example, a "z-up" quantum state: a state of spin in which a spin-½ particle is predisposed (with certainty) to be deflected in the positive z-direction if passed through this Stern-Gerlach device. Similar definitions are obviously available for "x-up" and "y-up."

Now we cannot simply define physical states into existence: it is an open question whether spin ½ particles can be disposed to react to a Stern-Gerlach device in this way. But experimentation settles that question because we are able to prepare beams with the relevant disposition. So the existence of such states, in this case, is settled empirically.

The *uniqueness* of these spin states, on the other hand, is quite a different matter. Is there *only one* physical state of spin that produces this disposition to be deflected in a given direction, or might there be many? As we have seen, given Albert's proposal for understanding the metaphysics of the quantum state (which starts from the standard mathematics rather than from the experimental facts), there are an infinitude of *different* possible quantum states that produce each of these dispositions. These states would differ by the (supposed physical analog) of the overall phase of the wave function. But because these supposed physical differences produce no observable differences in behavior, the standard supposition is that each of these spin states is the same: there is, for example, exactly one z-up state as we have defined it. Our job, at this point, is to invent *mathematical* representations of these states, rather than merely linguistic terms such as "z-up."

Once again, we are in some danger of confusing physical items with mathematical items, so a notation convention will come in handy. As we have seen, there exists a physical condition that disposes spin-½ particles to be deflected in the positive z-direction by a Stern-Gerlach apparatus, and we are further supposing that there is exactly one such physical condition with this property. We indicate this *physical* state with ceiling brackets, so the state in question is $\lceil z{-}up \rceil$. What we are seeking now is a perspicuous *mathematical* representation of $\lceil z{-}up \rceil$ and all the other possible physical spin states.

Here, our matrices come into play. Granting that there is only one $\lceil z{-}up \rceil$ quantum state, the usual thing that is said is that we arbitrarily choose the matrix $\begin{bmatrix} 1 \\ 0 \end{bmatrix}$ to represent it. $\lceil z{-}down \rceil$ is represented by $\begin{bmatrix} 0 \\ 1 \end{bmatrix}$, $\lceil x{-}up \rceil$ by $\frac{1}{\sqrt{2}}\begin{bmatrix} 1 \\ 1 \end{bmatrix}$, $\lceil x{-}down \rceil$ by $\frac{1}{\sqrt{2}}\begin{bmatrix} 1 \\ -1 \end{bmatrix}$, $\lceil y{-}up \rceil$ by $\frac{1}{\sqrt{2}}\begin{bmatrix} 1 \\ i \end{bmatrix}$, and $\lceil y{-}down \rceil$ by $\frac{1}{\sqrt{2}}\begin{bmatrix} 1 \\ -i \end{bmatrix}$. Any of these pairs of matrices forms an orthonormal basis for the vector space of 2×1 complex matrices. For example, an arbitrary matrix $\begin{bmatrix} a \\ b \end{bmatrix}$ can be written as $a\begin{bmatrix} 1 \\ 0 \end{bmatrix} + b\begin{bmatrix} 0 \\ 1 \end{bmatrix}$.

So if we are a bit sloppy, we might be tempted to write

$$\lceil x{-}up \rceil = \frac{1}{\sqrt{2}} \lceil z{-}up \rceil + \frac{1}{\sqrt{2}} \lceil z{-}down \rceil.$$

Carrying on our sloppy thinking, we might say that this last expression shows that the $\lceil x{-}up \rceil$ state is a superposition of two other spin states.

Let's see why the last few sentences were sloppy. If we accept the view that the overall phase of the *mathematical representation* of a quantum state has no physical significance, then the choice of $\begin{bmatrix} 1 \\ 0 \end{bmatrix}$ to represent $\lceil z{-}up \rceil$ was somewhat arbitrary: we could have equally well chosen $\begin{bmatrix} -1 \\ 0 \end{bmatrix}$ or $\begin{bmatrix} i \\ 0 \end{bmatrix}$ or in general $\begin{bmatrix} e^{i\theta} \\ 0 \end{bmatrix}$ for any value of θ. So relative to the choice of $\begin{bmatrix} 1 \\ 0 \end{bmatrix}$ to represent $\lceil z{-}up \rceil$, $\begin{bmatrix} 0 \\ 1 \end{bmatrix}$ to represent $\lceil z{-}down \rceil$, and $\frac{1}{\sqrt{2}}\begin{bmatrix} 1 \\ 1 \end{bmatrix}$ to represent $\lceil x{-}up \rceil$, we can write $\lceil x{-}up \rceil = \frac{1}{\sqrt{2}}\lceil z{-}up \rceil + \frac{1}{\sqrt{2}}$ $\lceil z{-}down \rceil$, but relative to some other choice of a mathematical representative the equation will be different. In short, although there is a well-defined operation of adding the *mathematical representations* of quantum states, and of multiplying them by complex numbers, there is no corresponding *physical* operation of "adding" the states themselves.

We could try to squeeze some of the arbitrariness out of the mathematical representations by choosing just one matrix to represent each spin state, such as the choices already made. But notice that once we do this, the mathematical representatives of the physical states no longer form a vector space. It is a requirement in a vector space, for example, that every vector $\boldsymbol{V}$ has an additive inverse $-\boldsymbol{V}$ such that $\boldsymbol{V} + (-\boldsymbol{V}) = \boldsymbol{0}$. In the vector space of 2×1 matrices, the inverse of $\begin{bmatrix} 1 \\ 0 \end{bmatrix}$ is $\begin{bmatrix} -1 \\ 0 \end{bmatrix}$. If we ban $\begin{bmatrix} -1 \\ 0 \end{bmatrix}$ from our set of mathematical representations, we can no longer regard the set of physical quantum states as a vector space.

The proposition that quantum states themselves (as opposed to their mathematical representatives) do not form a vector space, and a fortiori do not form a Hilbert space, is perhaps not terribly significant: the physical states of systems in *classical* mechanics do not form a vector space either. But as Abhay Ashtekar and Troy Schilling note, the linear structure of Hilbert space plays a central role in the standard presentations of quantum theory:

> While the classical framework is *geometric* and *non-linear*, the quantum description is intrinsically *algebraic* and *linear*. Indeed, the emphasis on the underlying linearity is so strong that none of the standard textbook postulates of quantum mechanics can be stated without reference to the linear structure of [Hilbert space]. (Ashtekar and Schilling 1998, p. 25)

Or, to take an example not discussed in the standard textbooks, one usual presentation of the measurement problem relies on the *linearity* of the evolution of the wave function: if the wave fuction $|\psi\rangle_0$ is mapped by the Schrödinger equation to $|\psi\rangle_1$ when the parameter t is increased from 0 to 1, and the wave function $|\phi\rangle_0$ is mapped by the Schrödinger equation to $|\phi\rangle_1$, then it follows from the linearity of the Schrödinger equation that $(|\psi\rangle_0 + |\phi\rangle_0)$ is mapped to $(|\psi\rangle_1 + |\phi\rangle_1)$. This mathematical fact clearly requires that the mathematical representations $|\psi\rangle_0$ and $|\phi\rangle_0$ have a well-defined sum. But one might well wonder exactly what this observation could tell us about the dynamics of quantum states if the quantum states themselves have no well-defined sum. More generally, if physical states themselves cannot be "added" or "multiplied by complex numbers," then one might well seek mathematical representatives of the states for which no such operations are defined.

Although vectors in a complex Hilbert space can be added and multiplied by complex numbers, *rays* in a complex Hilbert space cannot be. A natural suggestion is to try to formulate quantum mechanics directly in terms of the rays rather than vectors. That is, a natural suggestion is to formulate the mathematics of quantum mechanics in terms of *projective Hilbert space*. Such a change in the mathematics would eliminate some of the purely gauge degrees of freedom in the mathematical representation, and so would provide fewer opportunities for mathematical features of the representation to be mistaken for physical features of the thing represented.

It is not obvious how to formulate standard quantum mechanics directly in terms of projective Hilbert space. Ashtekar and Schilling (1998) provide some insights into how this can be done and how the resulting theory resembles classical mechanics more closely than one might have imagined. But my aim here is not to pursue the details of such a reformulation, it is simply to note the effect that such a reformulation could have on our basic ontological problem.

Albert (1996) starts with the standard mathematical formulation of quantum mechanics, which represents the quantum state by a vector in a Hilbert space. He asks after the simplest, most straightforward *ontological* account of the quantum state that would make evident why vectors in a complex Hilbert space should arise as representations. As we have seen, the simplest approach is to do one's best to reify every mathematical feature of the representation, that is, to postulate *physical* facts that correspond in some direct way with every *mathematical feature* of the representation. In Albert's case, this means postulating some unobservable physical degree of freedom that correspond to the mathematical phase, and also postulating some new *physical* principles that account for the use of only *normalized* vectors in the standard framework:

> And so (for example) the fact that the integral over the entirety of configuration [space] of the square of the amplitude of the universe's wave function is invariably equal to one is going to have to be thought of as not following analytically from the sorts of physical objects wave function[s] *are* (which it certainly cannot be), but as a *physical law*, or perhaps as an *initial condition*. (Albert 1996, p. 278)

Here, Albert takes not just the phase but also the absolute value of the amplitude of the wave function to reflect a physical feature of the quantum state. This is a very radical suggestion, hard to comprehend. For example, it could not possibly be a *law* that the integral of the squared amplitude of the electric field over all space be equal to one, because the numerical value of the integral depends on the gauge in which the field is measured. In any case, if we couch the theory in projective Hilbert space rather than in Hilbert space, none of these ontological suggestions even make sense.

The upshot of these considerations is clear: if one intends to try to read the physical ontology of a theory off of the mathematical structure used to present the theory, then one should give a great deal of consideration to alternative mathematical structures and the reasons for choosing one or another. The average physicist will not care much whether standard quantum mechanics is presented using Hilbert space or projective Hilbert space, but the average physicist is not in the business of giving clear ontological accounts. If we want to do better, then we have to become more sensitive to mathematical niceties from the outset.

In our test case, we now have two alternative mathematical structures: standard quantum mechanics formulated in terms of vectors in Hilbert space and a version of standard quantum mechanics formulated in terms of elements of projective Hilbert space. If we try to take each of these, as far as possible, "at face value," they yield different accounts of the physical degrees of freedom in a system. How should we decide which framework to use in our investigation?

The guidepost we have adverted to so far is epistemic: adopting Albert's understanding of the Hilbert space version commits us to the physical existence of degrees of freedom (e.g., the overall phase) that are not empirically accessible. Now I do not take empirical inaccessibility as a proof of physical nonexistence. For example, I am perfectly happy to consider the possibility of a foliation of space-time that plays a role in the fundamental dynamics but is not, because of the structure of that dynamics, open to empirical determination. But empirical inaccessibility certainly raises some legitimate concerns about a physical postulate.

One suspects that such an inaccessible feature could be purged from a physical ontology without damage to the remainder, yielding a more plausible theory. It is certainly a good heuristic to see if this can be done.

Once we focus on the question of empirical accessibility, what immediately stands out is how tremendously *inaccessible* the quantum state is. For example, we have theories according to which the quantum state always evolves deterministically and smoothly, such as Bohmian mechanics and the Everett theory, and theories according to which it suffers sudden, spontaneous, massive changes, such as GRW. Off-hand, one would expect such radically different dynamics to be easily distinguished empirically: why not just *look* at how the quantum state behaves? If a rival to Fourier had proposed an alternative to his smooth differential equations, insisting that the heat distribution in an object sometime underwent sudden massive changes, we would advise him to go into the lab and demonstrate such a remarkable phenomenon. Fourier's theory and this proposed rival could not long both survive as viable physical theories. The fact that both Bohmian mechanics and GRW are still on the table testifies to how distant the quantum state is from experimental data. As Bell said of Bohmian mechanics:

> Although Ψ is a real field it does not show up immediately in the results of a single "measurement," but only in the statistics of many such results. It is the de Broglie-Bohm variable X [i.e., the particle positions] that shows up immediately each time. That X rather than Ψ is historically called a "hidden" variable is a piece of historical silliness. (Bell 1987, pp. 162–163)

It is exactly because the quantum state is not directly observable that there can be so much latitude in picking a mathematical representation for it and in the dynamics ascribed to it. This should make us extremely cautious when we attempt to draw ontological morals from considerations of the mathematical representation.

So far, I have not even mentioned the most controversial aspect of Albert's (1996) position. Not only does the view developed there ascribe physical significance to the phase of the wave function, it interprets the mathematical space on which the wave function is defined as a direct representation of the fundamental physical space postulated by the theory. That is, according to Albert the mathematical item we denominate "configuration space" does not, in itself, have any connection to the configuration of anything. It is rather a fairly direct mathematical representation of a very, very high-dimensional *physical* space. As for the low-dimensional space-time that we are accustomed to thinking we live in, Albert calls it "somehow flatly illusory" (1996, p. 277) and a "mirage" (p. 279). The main burden of "Elementary Quantum Metaphysics" is to explain how such an illusion could have come about.

There is much to say about Albert's claim that the appearance of a low-dimensional space-time must be traced to the dynamics of a theory, and that the dynamics of the quantum state does not itself suggest that such a low-dimensional space-time is physically fundamental. I hope that the discussion of the much more pedestrian problem of the phase of the wave function (when quantum mechanics is formulated in terms of a Hilbert space) has served to cast doubt on the basic method of Albert's analysis. His desire is to take the mathematics at face value, insofar as possible, when proposing an ontology for quantum theory. But the mathematical framework itself (and especially the Hilbert space formulation) is not closely enough tied to empirical considerations to justify giving it so much deference.

There are many levels on which we can question the physical significance of some aspect of a mathematical formalism. At the most generic level, as Ashtekar and Schilling (1998) note, mathematical physics sometimes uses *algebraic* representations of physical states. That is, the representations admit of the usual operations characteristic of numbers: they can be added to one another, perhaps multiplied by one another, and multiplied by scalars. But it is typically obscure what sort of *physical* relation these mathematical operations could possibly represent. In the Hilbert space formalism, one can multiply any wave function by i to get a mathematically different wave function, but it would be very controversial to claim that there are two distinct quantum states for these wave functions to represent. Similarly, why think that there is any physical relation among physical states that corresponds to the mathematical operation of addition? This is especially vexing if the mathematical representatives form a vector space. By definition of a vector space, for each vector there must be an additive inverse that sums to the zero vector. Does the zero vector itself represent any possible physical state? In the case of spin, the matrix $\begin{bmatrix} 0 \\ 0 \end{bmatrix}$ is not taken to represent any possible spin state of a spin-½ particle. Further, the additive inverse of any matrix is taken to represent the same spin state as the matrix itself. Given these presuppositions, there cannot be any physical relation that corresponds to addition of the matrices: any physical state "added" to itself would have to be both represented by the zero matrix and by double the original matrix.

When Ashtekar and Schilling seek a *geometrical* representation of quantum states, in part they mean a mathematical representation for which these algebraic operations are not defined. On the one hand, this may make the mathematical manipulation of the representations more difficult: the linear structure of a vector space is a very convenient feature for solving equations. But if there are no physical analogs to the algebraic relations, the geometrical representation can have fewer gauge degrees of freedom. The way the mathematics corresponds to the physical ontology can then be made more transparent.

The burden of Albert (1996) is not to argue for a physical analog to the overall phase of the wave function. Albert would acknowledge that if such an analog exists it is unobservable, and he might well welcome the use of a projective representation from which the overall phase has been expunged. The main target of that paper was rather the representational significance of a different part of the mathematical machinery: the "configuration space," which serves as the domain of the complex functions that are the elements of the Hilbert space.

As the previous sentence testifies, the relevant part of the mathematical structure is fairly deeply embedded in the machinery. The algebraic approach begins with the idea that quantum states ought to be represented by elements of a complex vector space, in particular a complex Hilbert space. We have already seen that this choice seems to preclude from the beginning a one-to-one relation between quantum states and their mathematical representations. In standard quantum mechanics, each of the vectors in the Hilbert space is itself presented as a complex square-integrable function on another space. This high-dimensional mathematical space is commonly called "configuration space," and its mathematical structure is assumed to be isomorphic to the configuration space of a collection of classical point particles, which themselves exist in some common low-dimensional Riemannian space.

In classical physics the low-dimensional Riemannian space (recall this is still a *mathematical* object) is itself supposed to represent the geometrical structure of *physical* space. To take the obvious example, in classical physics one uses the mathematical space R^3 (the set of triples of real numbers (x, y, z)) together with

$$\text{the metric function } D((x_1, y_1, z_1),(x_2, y_2, z_2)) = \sqrt{\left((x_1 - x_2)^2 + (y_1 - y_2)^2 + (z_1 - z_2)^2\right)}$$

to represent three-dimensional Euclidean space E^3. The mathematical representation of E^3 already contains gauge degrees of freedom that reflect arbitrary decisions about the location of the "origin" (i.e., the point in *physical* space denoted by the triple of real numbers $(0, 0, 0)$); the orientation of the x-, y-, and z-axes; and the unit in which distance is to be measured. A triple of real numbers could only come to represent a point in physical space once these conventions have been settled.

Notice that we already have a *numerical* (and hence algebraic) representation of an inherently geometrical and nonalgebraic object. E^3 is not composed of numbers, and its elements (geometrical points) are not subject to arithmetical operations. Given two points p and q in E^3, it make no sense to ask for a point which is their "sum" or "difference," or what point one gets if one multiplies p by some scalar. There is no *geometrical* structure that corresponds to these algebraic operations. Of course, one can perform these algebraic operations on the *representations* of p and q: $(x_p, y_p, z_p) + (x_q, y_q, z_q) = (x_p + x_q, y_p + y_q, z_p + z_q)$ and $2 \times (x_p, y_p, z_p) = (2x_p, 2y_p, 2z_p)$. But which point in E^3 represented by any of these triples of real numbers is clearly gauge-dependent: given different coordinatizations of E^3, this supposed sum of p and q would be different, which is just another way of saying that there is no such sum at all. The algebraic structure of the representation of points in space corresponds to no geometrical structure in the space.

Given a coordinatization of E^3, we can indicate a point by means of a triple of real numbers. So one might naturally think the obvious way to indicate two points is by use of six real numbers, the first three providing coordinates of one point and the second three coordinates of the other. The obvious way to represent N points is by an ordered $3N$-tuple of numbers. The collection of all such $3N$-tuples forms a $3N$-dimensional vector space, commonly called the *configuration* space for N points in the base space (in this case, E^3). So we naturally generate a $3N$-dimensional vector space as a means of representing N points in the three-dimensional geometrical space. (Again, the algebraic structure of the space is an artifact of the coordinatization, and so does not correspond to anything physical.)

But, as Goldstein, Taylor, Tumulka, and Zanghì (2005) have pointed out, this usual mathematical representation of collections of N points in E^3 contains redundant representations. If, in a given coordinatization, the points p and q receive coordinates (0, 0, 0) and (1, 1, 1), respectively, then the pair of points $\{p, q\}$ can equally be represented by (0, 0, 0, 1, 1, 1) or by (1, 1, 1, 0, 0, 0) in the standard configuration space. One way to deal with this overabundance of mathematical representations is to collect them together into equivalence classes under permutation: one thereby would "quotient out" the redundant mathematical structure. The resulting space, which is (almost[4]) isomorphic to the set of all sets of N points in E^3, is called $^N R^3$. The motion of N point particles in E^3 would then correspond uniquely to the trajectory of a single point in $^N R^3$, instead of to a multiplicity of trajectories in R^{3N}. If one were designing a mathematical theory to represent the motion of N particles in E^3, an appropriate fitting of the mathematical machinery to the ontology would lead you first to $^N E^3$ (the set of sets of N points in E^3) and then, via a coordinatization of E^3, to $^N R^3$.

But $^N R^3$ is a much more complicated and messy mathematical object than R^{3N}. For example, it is much easier to write down differential equations that determine trajectories in R^{3N} than in $^N R^3$. So considerations of practicality might well suggest using R^{3N} as the mathematical representation of the physical situation, keeping the redundancy of the mathematics in mind. If *classical* mechanics had been developed with this much care about the mathematical niceties, spaces with the form of $^N R^3$ would have been given the name "configuration space" (because points in them more directly correspond to configurations of N points in E^3). R^{3N} would be relegated to the status of a convenient mathematical auxiliary whose relation to the physical situation is somewhat oblique.

----

[4] Why "almost"? Even after quotienting R^6 by permutations (in a two-particle theory), the resulting space would include, for example, a point corresponding to (0, 0, 0, 0, 0, 0), which represents both particles as occupying the "origin." But $^2 E^3$, the set of pairs of points drawn from E^3, has no corresponding element.

All of this mathematical detail bears directly on Albert's (1996) argument. He begins by considering an "N-dimensional classical-mechanical configuration space, in which a single world-particle is floating around" (p. 280). He goes on to ask how such a situation could come to have the *appearance* of a collection of particles in a lower-dimensional space. He begins with a world-particle governed by a free Hamiltonian, whose trajectory would therefore be a straight line in the configuration space, traversed at constant speed.

> And note that the trajectory of a world-particle like this one can patently contain *no suggestion whatever* as to whether we are dealing here with a single material particle moving freely in an N-dimensional physical space, or (say) N/3 distinct material particles moving freely in the *three*-dimensional physical space, or N distinct particles moving in a *one*-dimensional physical space. *Nothing* about a trajectory like that (to put it slightly differently) can make it natural or make it plausible or make it reasonable or make it simple or make it elegant or make it any other desirable thing to suppose that any *one* of those possibilities, as opposed to any of the *others* I mentioned, or any one of the others I *did*-*n't* mention, actually obtains. (Albert 1996, p. 280)

Albert concludes that the appearance of a low-dimensional space must depend on the existence of something other than a free Hamiltonian (i.e., on what are normally called interaction terms in the Hamiltonian), and he goes on to consider just what sort of Hamiltonian might make the postulation of a low-dimensional physical space plausible.

But if the mathematics in which classical mechanics is couched had been developed in a more philosophically sensitive way, Albert's puzzle would never get off the ground. If one uses R^{MN} to represent the configuration space of N particles in an M-dimensional Euclidean physical space, then the configuration spaces of one particle in a six-dimensional space, two particles in a three-dimensional space, three particles in a two-dimensional space, and six particles in a one-dimensional space will all be represented by R^6. But if one uses $^N R^M$, then these configuration spaces are all mathematically distinct from the get-go. *Even before* any trajectory has been specified, the number of particles and the geometry of the physical space they inhabit can be recovered from the mathematical structure of the configuration space. Worries about the Hamiltonian never even arise.

The use of $^N R^M$ in classical mechanics would also have avoided some obvious problems in the standard presentations of the theory. It is usual to say (although very problematic in any case) that the "observables" of classical mechanics correspond mathematically to the smooth real-valued functions on the phase space of the system. But if we use R^{MN} to represent the configuration space, there will be "observables" on the two-particle phase space which take different values at (0, 0, 0, 1, 1, 1; 0, 0, 0, 0, 0, 0) in the phase space than at (1, 1, 1, 0, 0, 0; 0, 0, 0, 0, 0, 0), even when the particles have all the same intrinsic properties. Intuitively, each of these points in phase space should represent the same physical situation: one particle at the point labeled (0, 0, 0), another at the point labeled (1, 1, 1), each with zero momentum. If we take "observable" seriously, this would mean that the two situations could be observationally distinguished. But if the particles are qualitatively identical, no observation could distinguish these situations. So either the observables have to be restricted in some way (e.g., by requiring that they take the same value at these two points), or, more reasonably, the phase space needs to be reduced. For the only *justification* for the restriction on the observables is that the two points in the phase space represent the same physical situation—in which case there ought not to be two points in the phase space to begin with.

Note that as with the overall phase of the wave function, considerations of observability are brought into play when crafting a mathematical representation of the physical state. Since the mathematical representation may then be used to suggest accounts of the physical ontology, epistemology and considerations of observability will influence the metaphysics. The circle would close when the physical ontology and dynamics are used to model the observer and explain which features of a physical system can influence an observer in the right way to be reliably observed.

The approach followed in Albert (1996) jumps into the middle of this dialectic. A mathematical formalism is accepted, and an ontology crafted to mirror it, without first considering how the formalism was developed in the first place, and hence which features of the mathematics ought to be taken seriously. The end result is the postulation of a physical space with a geometrical structure isomorphic to the "configuration" space and a set of physical properties isomorphic to the amplitude and phase of the wave function. But if one steps back to examine where the mathematical formalism came from in the first place, and the parts of it that generate the observable behavior that is used to test the theory, one would not be likely to postulate this sort of physical structure.

Let's try to sketch a rough account of the development of a physical theory. Any physical theory begins with prima facie data: some sort of physical behavior that is regarded as easily observable and hence fit to serve as evidence against which the theory can be tested. Let's call this prima facie data the *Primary Observables* of the theory. For some (purported) physical fact to be included among the Primary Observables, one must be able to provide some account of how it can be observed. Note: postulation of something as a Primary Observable is itself a sort of theoretical act and may turn out to be wrong. But methodologically, it is desirable for the Primary Observables to be couched in a theoretically neutral language, that is, for the Primary Observables to be the sorts of things that competing theories will tend to hold in common.

A couple of examples. In the Scholium to the Definitions in *Principia*, Newton distinguishes absolute, true, and mathematical time, space, place, and motion from relative, common, and apparent time, space, place, and motion. He acknowledges that the relative, common, and apparent versions are the Primary Observables in his theory. That is, he acknowledges that no simple experiment or observation can establish the absolute motion of an object, but he accepts that relative motion is easy to determine. The relative motions are the *phenomena*, the apparent facts, that form the data of the theory. Or consider thermodynamics. The *distribution of heat* in (certain sorts of) macroscopic objects is one of the Primary Observables for Fourier. A typical passage:

> When a metal bar is exposed at one end to the constant action of a source of heat, and every point of it has attained its highest temperature, the system of fixed temperatures corresponds exactly to a table of logarithms; the numbers are the elevations of thermometers placed at the different points, and the logarithms are the distances of these points from the source. (Fourier 1952, p. 183)

Fourier assumes that this is the sort of physical fact one can check directly and unproblematically. It is therefore the sort of physical fact that any acceptable theory of heat must account for. Or again, the pressures, volumes, and temperatures of boxes of gases are (within observational limits) Primary Observables. The weights of the various ingredients and products of combustion and fermentation are Primary Observables for Lavoisier: these are the data on which he builds his theory. The notion of a Primary Observable evidently has an epistemic aspect: a (purported) physical characteristic is only ascribed this status if we think we know how to determine it experimentally. But the facts themselves make no reference to observation: the relative motions of objects, or the distribution of heat in a metal bar, or the proportion of oxygen and hydrogen that are the products of hydrolysis are what they are independently of whether anyone looks.

A physical theory postulates an ontology: a collection of items taken to be physically real. The laws of the theory are couched in terms of this ontology. The Primary Observables must be (according to the theory) determined by the behavior of this ontology: otherwise, the theory will not make contact with its data. But it may also turn out that the Primary Observables are (postulated to be) a function of a *proper subset* of the ontology of a theory. That is, the behavior of just that subset determines the data, and hence the empirical consequences of the theory would be the same even supposing the account of the rest of the ontology were altered or rejected, provided that the behavior of this subset were unaffected. Intentionally modeling this choice on the terminology of Allori, Goldstein, Tumulka, and Zanghì (2008), let's call that part of the ontology of a theory that determines the Primary Observables the *Primary Ontology*.[5]

---

[5] I do not adopt their terminology because I want the relation between what I am defining and what they are defining to be open. Clearly, the notions are in the same ballpark. I don't want to devote time to a more minute comparison.

The value of the Primary Observables is, by definition, some function of the state of the Primary Ontology, but there still may be a rather wide gap between the sorts of things that fall in each category. For example, the Primary Observables are all macroscopic quantities, but the Primary Ontology may well be microscopic entities. In this case, the Primary Observables concern the gross behavior of large ensembles of elements of the Primary Ontology. Our only constraints are that (1) the Primary Observables should supervene on the state of the Primary Ontology and (2) the parts of the physical ontology deemed Primary should play a role in determining the Primary Observables.

Because the Primary Ontology is defined with respect to the Primary Observables, the division of the whole Ontology into Primary and (let us say) Secondary has an epistemic cast. This is not a distinction into different *kinds of existence*. It is a distinction that should track which parts of a theory, according to the theory itself, are more directly and unproblematically related to empirical data and which are more remote from empirical data, and hence more speculative. Evidently, not everything the theory postulates about the Primary Ontology is accessible to direct observation: there may be plenty of speculation there as well. Some examples will help make this clear.

Suppose that we are interested in thermodynamics. Then the Primary Observables will include, as Fourier illustrates, the (macroscopic) distribution of heat (or temperature) in certain objects. Now a thermodynamic theory may contain a *metaphysical analysis* of heat: it may proclaim, for example, that heat is a substance (caloric theory) or that heat is just the motion of molecules (kinetic theory). In the first case, caloric is part of the Primary Ontology of the theory, and in the second the microscopic motion of molecules is part of the Primary Ontology. For in each case, the distribution of these things accounts for the Primary Observable, namely, the distribution of heat. A theory may go on to postulate further bits of ontology in the service of accounting for the behavior of caloric or the transmission of microscopic motion among molecules. But the nature and behavior of these further things would be part of the Secondary Ontology, because that behavior is *screened off* from the data by the Primary Ontology. That is, if we imagine keeping the behavior of the Primary Ontology fixed but altering the behavior of the Secondary Ontology, the data would remain the same.

Notice that classifying caloric among the Primary Ontology of the caloric theory of heat and microscopic molecular motion among the Primary Ontology of the kinetic theory does not imply one can directly observe the existence of either of these or their status as providing the fundamental nature of heat. *That* heat is caloric is not directly observable, but *if* heat is caloric (as the theory asserts), then the distribution of caloric in an object is a Primary Observable. If the very *existence* of the entire ontology postulated by a theory is classed among the Primary Observables, then the theory itself is called a *phenomenological* theory, that is, a theory that does not postulate the existence of anything beyond what is accepted as observed. Both the caloric and kinetic theories postulate an unobservable physical nature of heat, and so are not phenomenological. But the distribution and motion of heat (without further claims about its nature) was considered directly observable. A phenomenological theory seeks only a description of this observable behavior. Fourier's own theory largely falls here: he thought he could just observe the distribution of heat on the rod, and he offered no particular metaphysical account of what heat is. Rather, he sought a simple mathematical description of the observed heat flow.

In the case of Newton's theory, the Primary Observables are the relative positions and motions of things. Any theory that agrees with Newton's on these will agree on all the empirical data. Newton postulates the existence of matter in absolute space and time. This is part of his Primary Ontology because the Primary Observables are analyzed in terms of it: the relative motions are just the differences of the absolute motions. So absolute motion belongs in his Primary Ontology even though it is not among the Primary Observables. Newton's theory is not phenomenological because his fundamental laws are not couched in terms of the Primary Observables: he needs absolute motion to state his laws. So he is (as he is aware) *immediately* committed to the existence of not-directly-observable facts in his explanation of the observable facts. Absolute motion and absolute space and time, although not directly observable in Newton's theory, cannot be classed as Secondary Ontology of the theory because the Primary Observables are ontologically explicated in terms of them.

All of this taxonomy is doubtless squishy and somewhat obscure. One could discover problem cases in which the application of the taxonomy seems arbitrary. Nonetheless, I think we can apply it pretty well. For example, consider neutrinos. No one would claim that the existence or trajectory of a neutrino was among the Primary Observables in physics in the 1930s. Nor were they among the Primary Ontology *even of theories that postulated them*: the *observable* behavior of laboratory equipment was not taken to be *constituted* by the behavior of neutrinos. The ontology of the theory could be divided into neutrinos and "normal" matter, and the data of the theory were directly a function of the behavior of the normal matter. This is not a *physical* distinction: in an obvious sense, the neutrinos were postulated to be just another sort of particle, like the electron and the neutron. But the *confirmation theoretic* situation was that the behavior of the normal matter screened off the data from the behavior of the neutrinos. Our empirical access to neutrinos is mediated by the normal matter in a way that requires detailed analysis of causal interaction and theoretical postulation. In contrast, the relative motions of things arise from their absolute motions (according to Newton) in a straightforward analytical way.

The point of introducing this taxonomy is methodological: one ought to be much more tentative and cautious about the nature and behavior of the Secondary Ontology of a theory than about its Primary Ontology. The Secondary Ontology is, by definition, more remote from our empirical access. Indeed, one might rightly wonder what sort of evidence we can have for the existence, behavior, and nature of the Secondary Ontology at all. On the flip side of the same coin, one ought to be more flexible and creative when considering the nature of the Secondary Ontology because it only needs to fulfill a functional role with respect to the Primary Ontology in contributing to the empirical content of a theory.

Let's try to apply these lessons to quantum mechanics, or rather, to a few different theories that use the mathematical formalism of quantum theory. The easiest case is Bohmian mechanics. In Bohmian mechanics, as in Newtonian mechanics, one may take the Primary Observables to be the relative locations of macroscopic collections of matter. In the usual idealization, we imagine a laboratory apparatus to be a box with a big pointer on it exactly because we think we can reliably tell, just by looking, which way the pointer ends up pointing with respect to the scale on the device. The Primary Ontology is the matter that is postulated to constitute, among other things, these macroscopic collections. The exact structure of the Primary Ontology is not observable. Bohmian mechanics takes it to be particles, perhaps even point particles, although the theory may be agnostic on that detail. More particularly, the Primary Ontology (as we have defined it) will consist in the sorts of particles that are postulated to make up objects that are Primarily Observable. Bell, in his discussion of choices for local beables, selects fermion number density as a local beable for just this reason:

> Not all "observables" can be given beable status, for they do not all have simultaneous eigenvalues, i.e. do not all commute. It is important to realize therefore that most of these "observables" are entirely redundant. What is essential is to be able to define the positions of things, including the positions of instrument pointers or (the modern equivalent) of ink on computer output....
>
> We fall back then on a second choice—fermion number density. The distribution of fermion number in the world certainly includes the positions of instruments, instrument pointers, ink on paper ... and much much more. (Bell 1987, p. 175)

Although Bell did not use anything like our taxonomy, he must have had some nearby considerations in mind. The *exact* distribution of fermions, which is what I have called his Primary Ontology, is clearly not a directly observable matter. Indeed, Bell apparently feels confident that he can leave all bosons (including photons) entirely out of account and not run into any difficulties. Even the *existence* of fermions is not a Primary Observable: one cannot tell, just by looking, that pointers and ink are made of fermions rather than something else. But the distribution of fermions, if we postulate their existence and the role they play in composing perceptible objects, is clearly sufficient to account for all data because, given such a posultation, the Primary Observables are all functions of the fermion distribution. Even if he does not explicitly mention it, epistemic motivations influence Bell's choice of local beables.

Our little structural account of theories also explains why, in a nonradical context, there should be particular concern about the local beables of a theory: in a nonradical context, all of the Primary Observables are taken to be local matters of fact. What one can observe, in the first place, are physical conditions that obtain in certain restricted regions of space and time: in a particular lab, for example, on a particular day. There are obviously also simple conjunctions of these local observations, such as a compendium of observations amassed over the course of a year, but these are just logical consequences of a set of more spatiotemporally restricted observations. If we accept this at face value (which requires accepting the *existence* of space-time at face value, something Albert is cajoling us not to do), then the Primary Ontology of a theory must *also* be local matters of physical fact. If we keep the local ontology of a theory fixed (in a certain situation) and change only the nonlocal ontology, the Primary Observables will be unchanged. If a theory has no local ontology, then it is hard to see how it can give rise to local observable facts. Our understanding of the Primary Observables would have to be radically different from any historical precedent in such a case.

Every standard (i.e., non-Albertian) version of Bohmian mechanics—including, as Bell says, quantum field theory—postulates some local beables in a low-dimensional space-time and postulates that the Primary Observables are functions of those local beables. In our terms, then, the Primary Ontology of such a theory is uniformly local. The quantum state in such a theory must be Secondary Ontology because the quantum state is not local. Even in Albert's nonstandard version of Bohmian mechanics, with a *single* particle moving in a very high-dimensional physical space, the Primary Ontology is just the single particle: the "appearances" are supposed to be a function of how the *particle* moves, not how the quantum state evolves. Even in this nonstandard setting, it remains that no experiment can *directly* reveal the quantum state of any system: our only clues to the quantum state are inferences from the behavior of the Primary Ontology.

Let's leave Albert's radical suggestion aside for the moment. In the standard versions of the theory, there is a local ontology (particles, say) in a low-dimensional space-time, organized to form pointers on boxes, patterns of ink on paper, and so on. The quantum state is introduced into the theory as part of the physical account of how the local beables come to be arranged as they are, and hence why the Primary Observables come out as they do. That is, the quantum state is introduced as part of Secondary Ontology while proposing a physical account of why the Primary Ontology behaves as it does. Even more precisely, the quantum state is introduced as part of a physical account of why the configuration of local beables evolves as it does.

Notice that at this point we have not said a single sentence about the *metaphysical nature* of the quantum state, except to say that it is not part of the

Primary Ontology. Here are three reasons to assert that it is not part of the Primary Ontology: (1) the direct observation that no experiment can *reveal* or *determine* the exact quantum state of a given system, (2) the related circumstance that the disposition of the local ontology screens off the quantum state from the data, and (3) the slightly more circuitous observation that the Primary Ontology should all be local because the Primary Observables are local, and the quantum state is not local. So all we really have postulated about the quantum state is that our epistemic access to it is *indirect*: ultimately, we only believe anything about the quantum state because we believe something else about some other part of the ontology of the theory, in particular something else about the behavior of the Primary Ontology.

These observations tell us nothing at all about the quantum state per se, but they do suggest something about what form the *mathematical representation* of the quantum state is likely to take. If our epistemic access to the quantum state is mediated through the behavior of the Primary Ontology—in particular if the only thing we know about the quantum state is how it influences the configuration of the Primary Ontology—then the mathematical representation of the quantum state ought be designed to facilitate predictions about how that configuration will (or might) evolve. The simplest mathematical representation of how a configuration might evolve is a *vector field* on the configuration space of a system. The simplest mathematical way to specify a vector field on a (mathematical) space is to take some sort of gradient of a scalar function on that space. So if the quantum state is the *only* physical entity (beside the present configuration itself) that determines how the configuration of a system will evolve, we can expect a priori that a scalar function on the configuration space of a system can serve as a convenient mathematical representation of the quantum state.

The foregoing paragraph contains an important condition: the evolution of the configuration of a system should be determined only by its present configuration and some other thing, which we call the quantum state. That condition might not hold. In classical mechanics, the evolution of a system is determined by the configuration and first time derivative of the configuration, together with other physical characteristics (masses, charges, etc.), which we might call the *classical state*. So the dynamics of a system like that can be represented by a function on the *phase space* (rather than the configuration space) of a system. The classical state should somehow determine a velocity field on the phase space, and again, a convenient way to represent this would be via some scalar function on the phase space. In classical mechanics, one such scalar function is the Hamiltonian function: any Hamiltonian function defined on the phase space of a system gives rise to a Hamiltonian vector field and Hamiltonian flow, which represents the evolution of the system in time. Although the particular mathematics may be complicated, the general idea that such a function should exist is understandable: the Hamiltonian function encodes all the relevant facts (forces, etc.) about the dynamics of a system save the positions and momentums of the particles. So specifying the positions and momentums should yield a trajectory for the system, represented by a trajectory in phase space. The fact that the relevant physical facts about a system can be conveniently summarized by a Hamiltonian function on the phase space tells us nothing at all about the metaphysical nature of those facts, but only reflects the generic situation that the positions and momentums of the particles, together with all the other relevant physical facts, determine how the systems evolves.

By parity of reasoning, the fact that the quantum state in Bohmian mechanics can conveniently be represented by a scalar field on the configuration space of the Primary Ontology reflects nothing more than the fact that the dynamics of the theory is first-order in time. Given the present configuration of the Primary Ontology, together with some other physical facts, the evolution of the system is determined. So the additional physical facts, whatever they are, determine a vector field on the configuration space. A convenient way to indicate a vector field on a mathematical space is to put a scalar field on the space and take a gradient (or something like a gradient).

Indeed, reflection on the way one derives a vector field from the mathematical representation of the quantum state (or, in classical mechanics, the classical state) immediately reveals that some of the mathematical degrees of freedom in the description are gauge, so one has no reason to think there is any corresponding physical degree of freedom in the state itself. For example, if one takes the simple gradient of a scalar function to get a vector field, then adding a constant to the scalar field will not change the vector field, and such an addition should be regarded as merely a change of gauge. Because the way the vector field in Bohmian mechanics is derived from the representation of the quantum state is more complicated, the gauge degrees of freedom in the mathematics are more extensive. Stated in terms of ψ, the guidance equation is

$$\frac{d\mathbf{Q}_j}{dt} = \frac{\hbar}{m_j}\text{Im}\frac{\psi^* \nabla_j \psi}{\psi^* \psi}(\mathbf{Q}_1,...,\mathbf{Q}_N).$$

It is clear that multiplication of the wave function ψ by any constant will not change $d\mathbf{Q}_j/dt$, so both the overall phase and overall scale of ψ are gauge degrees of freedom.[6] Contrary to Albert's claim that the use of normalized wave functions should reflect a physical law or a special initial condition, we see that changing the scale of the amplitude of the wave function makes no physical difference at all in the effect of the wave function on the Primary Ontology, and hence on the Primary Observables. We can therefore have no empirical reason

---

[6] For more detail on the guidance equation, see the contribution in this volume by Goldstein and Zanghì.

to think that the overall phase or scale of the wave function correspond to any physical fact.

More profoundly, we have also seen why, given that our epistemic access to the quantum state is mediated through the behavior of the Primary Ontology via a first-order dynamic equation, it is unsurprising that the quantum state might conveniently be represented by a scalar function on the configuration space of the system. Albert's initial thought was that taking the quantum state to be real (in the appropriate sense) suggests that it be understood as a field on a high-dimensional physical space. But that thought gains plausibility only if we (1) elide the distinction between the quantum state and its mathematical representation, (2) use the Hilbert space representation of the quantum state as a vector rather than some other representation, and (3) ignore the questions raised by the unobservability of the overall phase of the wave function, which suggest that it is merely gauge. If we are going to take the mathematical form of a representation as a guide to the metaphysical structure of the entity represented, then we need to consider carefully how that representation came to be in use, what alternatives there might be, and what conventions are in play in using algebraic objects as representations of physical entities.

Starting from the Bohmian Primary Ontology (particles in a low-dimensional space), together with the fact that the theory is deterministic and first-order in time, we have seen why a scalar field on the configuration space might naturally arise as a representation of whatever-else-there-is (beyond the configuration) that determines the evolution of the physical state. If we fold whatever-else-there-is into something called the quantum state, then the mathematical representation of the quantum state gives us no particular clue into its metaphysical status. The quantum state, on this reading is, of course, *physically real*. It plays a role in determining how things evolve, and so it is not somehow just a matter of (say) our knowledge of the state of the universe. But the particular mathematical form of the representation does not tell us what it is.

Nor, as we have seen, do we even have much of a grasp on what it *does*. The viability of both collapse and noncollapse theories indicates that even the most basic features of the dynamics of the quantum state are beyond any immediate experimental check. As is pointed out in Dürr, Goldstein, and Zanghì (1997) and Goldstein and Teufel (2001), in a Bohmian theory it is even an open question whether the quantum state of the universe changes *at all*.

Let's review how that comes about. First of all, in an obvious sense there is really only one quantum state: the quantum state of the entire universe. If the quantum state of the universe and the configuration of the universe determine the evolution of the universe, then they also determine the evolution of all subsystems of the universe. There is no room for any more independent ontology.

In a similar spirit, in classical mechanics the entire Secondary Ontology (including the various "forces") is coded up in the Hamiltonian of the universe. One can, of course, ascribe a Hamiltonian to certain subsystems of the universe, but it is logically derivative from the universal Hamiltonian and the environment of the subsystem. Notice that although one may use a *time-varying* Hamiltonian for a subsystem, the variation in time is really a reflection of the variation in the environment of the subsystem and the coupling of the subsystem to the environment. For example, if we put a subsystem in a slowly changing magnetic field, then we may analyze its behavior using a time-dependent Hamiltonian, but if we expand our scope to include the external source of the field, we can use a time-independent Hamiltonian. The natural expectation is that if we were to expand our scope to include the entire universe, then the Hamiltonian would be as constant as the laws of nature themselves.

Just as the use of time-varying Hamiltonians for subsystems in classical mechanics does not suggest that the universal Hamiltonian changes with time, so the use of time-varying wave functions in describing subsystems need not imply that the universal wave function (and hence the universal quantum state) varies in time. In Bohmian mechanics, at a universal level there is no "collapse of the wave function": the quantum state does not undergo any nonlinear changes. Nonetheless, the wave function *used to describe a subsystem* can undergo a non-linear change, which is a mathematical consequence of the smooth evolution of the Primary Ontology, the smooth evolution of the (universal) quantum state, and the way the wave function of a substem is defined from these. What this suggests is that even the Schrödinger evolution of the wave function of a subsystem might be explicable in the same way: the universal quantum state *does not change at all*, but the wave function of a certain kind of subsystem nonetheless obeys Schrödinger's equation.

This possibility has been discussed in the work of Dürr, Goldstein, Teufel, and Zanghì already cited. If the universal quantum state turns out to be static, then its metaphysical characteristics would come to resemble those we attribute to *laws* rather than those of fields or material entities. If the mathematical representation of such a universal quantum state were strikingly simple the analogy would be nearly perfect. It speaks volumes about the tenuousness of our epistemic access to the quantum state that such a possibility is consistent, as far as we know, with all our empirical data.

As long as we agree that this account might be correct, we agree that we do not know enough about the quantum state to determine even the most generic features of its behavior. But then we are in no position to make reliable pronouncements about what it is. We do not even know the right general ontological category in which to put it. Indeed, there is no reason to believe that any theorizing or speculation on the nature of the physical world that took place before the advent of quantum theory would have hit on the right ontological category for the quantum state: because it is so hidden, there would have been nothing relevant to speculate about. Whether one finds the possibility invigorating or disheartening, the best ontological category for the quantum state might simply be the category *Quantum State*, just as the right ontological category for a classical field is *Field*, not "stress in a medium" or "collection of particles."

This is not to say that we cannot try to characterize the quantum state: we would like to know, for example, how it influences the Primary Ontology and whether the universal quantum state changes with time. Different theories may present different answers to these questions (as well as the question of what the Primary Ontology is). We would like some sort of account of why the *mathematical representation* of the quantum state takes the form it does in any theory.

Albert's (1996) suggestion is that the mathematical representation of the quantum state is a rather transparent indication of its metaphysical structure. The ontological price he has to pay for this view is to elevate configuration space into a primitive, free-standing physical space (not a configuration space at all) and accept the physical significance of mathematical degrees of freedom that do not give rise to any observable differences. The Bohmian explanation runs quite differently: the mathematical configuration space really is a configuration space, whose points represent configurations of local beables in a low-dimensional space-time; the mathematical form of the representation of the quantum state is such as to make it easy to represent a flow on configuration space (and hence the dynamics of the Primary Ontology) as a function of the configuration and the quantum state. Different mathematical representations that generate the same flow are postulated to represent the same state, and the mathematical differences to merely be gauge. One can pursue, in this project, the development of a new mathematical formalism from which the gauge degrees of freedom have been purged.

There are, of course, other theories as well. Some of them (such as a GRW theory with a matter density ontology) appear to have a harder time explaining why the quantum state should be represented as it is: the configuration space of that Primary Ontology is not the configuration space of quantum mechanics. Of course, insofar as the observable world behaves as if it were made up of particles, there is an explanation of why *representing* it as made of particles should be empirically convenient. A theory without a particle ontology might be able to account for the phenomenological appearance of particle-like behavior (indeed, it has to, if it is to be empirically adequate). But such a story is of necessity somehow more convoluted than a theory that postulates particles, and we need to understand how the convolutions go.

The main point to bear in mind is that studying the mathematics in which a theory is couched is not the royal road to grasping its ontology. Mathematical convenience arises in many ways. One way is rooted in the mathematical (and in particular the geometrical) structure of the physical ontology itself. In such a case, the mathematical representation might, in an appropriate sense, be isomorphic to the physical object it represents. But most mathematical representation is more oblique. The mathematical objects acquire algebraic and numerical properties that the physical objects do not have; there are purely gauge degrees of freedom in the mathematics. Complicated physical states of affairs (such as the configuration of many particles in a low-dimensional space) are represented by mathematical objects with a different structure (such as a single point in a high-dimensional state). The nonlocality inherent in quantum theory entails that taking a common low-dimensional space-time as the arena of the Primary Ontology will force a more unfamiliar mathematical representation of the quantum state. Albert offers us a different calculus: abandon the low-dimensional space and build an ontology to match the mathematics. This is not the only way, and not the most natural way to proceed. If we take considerations of observability into account, we are naturally led to regard certain mathematical features of the wave function as not representing any characteristic of the quantum state, contrary to Albert's suggestion. Starting with a Primary Ontology like that of Bohmian mechanics yields a fairly straightforward account of why the wave function, the mathematical representation of the quantum state, might take the form it does. The ultimate questions about the dynamics and the fundamental nature of the quantum state, though, remain more elusive.

References

Albert, D. 1996. Elementary Quantum Metaphysics. In *Bohmian Mechanics and Quantum Theory: An Appraisal*, ed. J. Cushing, A. Fine, and S. Goldstein, Dordrecht: Kluwer, 277–284.

Allori, V., S. Goldstein, R. Tumulka, and N. Zanghì. 2008. On the Common Structure of Bohmian Mechanics and the Ghirardi-Rimini-Weber Theory. *British Journal for the Philosophy of Science* 59:353–89.

Ashtekar, A., and T. Schilling. 1998. Geometrical Formulation of Quantum Mechanics. In *On Einstein's Path*, ed. A. Harvery, New York: Springer-Verlag, 23–63.

Bell, J. S. 1987. *Speakable and Unspeakable in Quantum Mechanics*. Cambridge: Cambridge University Press.

Dürr, D., S. Goldstein, and N. Zanghì. 1997. Bohmian Mechanics and the Meaning of the Wavefunction. In *Experimental Metaphysics—Quantum Mechanical Studies for Abner Shimony, Volume 1*, ed. R. S. Cohen, M. Horne, and J. Stachel, Boston Studies in the Philosophy of Science 193, Dordrecht: Kluwer, 25–38.

Faraday, M. 1952. Experimental Researches in Electricity. In *Great Books of the Western World* vol. 45, Chicago: Encyclopædia Brittanica, 253–866.

Fourier, J. 1952. Analytical Theory of Heat. In *Great Books of the Western World* vol. 45, Chicago: Encyclopædia Brittanica, 161–252.

Goldstein, S., J. Taylor, R. Tumulka, and N. Zanghì .2005. Are All Particles Identical? *Journal of Physics A: Mathematical and Theoretical* 38:1567–76.

Goldstein, S., and S. Teufel. 2001. Quantum Spacetime without Observers: Ontological Clarity and the Conceptual Foundations of Quantum Gravity. In *Physics Meets Philosophy at the Planck Scale*, ed. C. Callender and N. Huggett, Cambridge: Cambridge University Press, 275–89.

Lewis, P. 2004. Life in Configuration Space. *British Journal for the Philosophy of Science* 55:713–29.

Maudlin, T. 2002. Thoroughly Muddled McTaggart or How to Abuse Gauge Freedom to Generate Metaphysical Monstrosities. *Philosopher's Imprint* 2 (4). Available online at http://www.philosophersimprint.org

7

Against 3*N*-Dimensional Space*

BRADLEY MONTON

1 Quantum Mechanics Is False

Question 1: How many dimensions does space have? I maintain that the answer is "three." (I recognize the possibility, though, that we live in a three-dimensional hypersurface embedded in a higher-dimensional space; I'll set aside that possibility for the purposes of this chapter.) Why is the answer "three"? I have more to say about this later, but the short version of my argument is that our everyday commonsense constant experience is such that we're living in three spatial dimensions, and nothing from our experience provides powerful enough reason to give up that prima facie obvious epistemic starting point. (My foils, as you presumably know from reading this volume, are those such as David Albert [1996] who hold that actually space is 3*N*-dimensional, where *N* is the number of particles [falsely] thought to exist in [nonexistent] three-dimensional space.)

Question 2: How many dimensions does space have, according to quantum mechanics? If quantum mechanics were a true theory of the world, then the answer to Question 2 would be the same as the answer to Question 1. But quantum mechanics is not true, so the answers need not be the same.

Why is quantum mechanics false? Well, our two most fundamental worked-out physical theories, quantum mechanics and general relativity, are incompatible, and the evidence in favor of general relativity suggests that quantum mechanics is false. For example, some of the evidence for general relativity involves experiments done with precise clocks; these experiments show that clocks in stronger gravitational fields run slow compared to clocks in weaker gravitational fields (see, for example, Hafele and Keating 1972a, 1972b). According to quantum mechanics, ideal clocks run at the same rate regardless of the strength of the gravitational field affecting them. Quantum mechanics makes predictions at variance with experiment, so quantum mechanics is false. (I recognize, for the record, that this argument is not definitive; arguments in science typically

* Thanks to David Albert, Alyssa Ney, and Ted Sider for helpful comments.

aren't. It could be auxiliary hypotheses that are false, not quantum mechanics. But I don't know of any plausible proposals for such false auxiliary hypotheses, so I assume that it's really quantum mechanics that is false.)

There are attempts by physicists to come up with a new theory that will replace both quantum mechanics and general relativity—yielding prototheories like loop quantum gravity, string theory, and M theory—but that project is very much ongoing, without clear results yet. If we are going to do physics-based metaphysics, it would be nice if we could base our metaphysics on a true fundamental physical theory, or at least on a fundamental physical theory that we had solid epistemic reason to take to be true. Sadly, we don't have such a theory. But one benefit for philosophers is that this makes the project of attempting to engage in physics-based metaphysics much more philosophically interesting. (For more of my thoughts on this, see the last section of this chapter, and for even more, see Monton 2010.)

2 Bohr, Schrödinger, and 3N-Dimensional Space

Let's focus on Question 2: how many dimensions does space have, according to quantum mechanics? To answer this question, it is helpful to step back and ask a more basic one: how does one determine the ontological content of a physical theory? Well, if we can (for whatever reason) presuppose that the theory is true, then the ultimate arbiter of the ontological content of the theory is reality itself. But for false physical theories, that presupposition is inappropriate. We cannot, for example, presuppose that Aristotelian physical theory is true—we wouldn't be correctly understanding the content of Aristotelian physical theory. Similarly, we can't presuppose that quantum mechanics is true. If we were to do so, we would conclude that quantum mechanics correctly predicts that clocks in stronger gravitational fields run slower, but quantum mechanics clearly makes no such prediction.

So how do we determine the content of, say, Aristotelian physics? One prima facie promising answer is: "we read Aristotle." What happens if we apply the analogous answer to the case of quantum mechanics? Quantum mechanics did not have a single developer, but Bohr and Schrödinger were two central figures, so let's look briefly at what they thought about the ontological status of quantum mechanics.

Bohr's writings on how to interpret quantum mechanics are notoriously unclear; Bohr himself is open to interpretation. One standard interpretation of what he says is that quantum mechanics cannot be used to describe the world, only the results of a given experimental arrangement. For example, he writes:

> there can be no question of any unambiguous interpretation of the symbols of quantum mechanics other than that embodied in the well-known rules which allow to predict the results to be obtained by a given

experimental arrangement described in a totally classical way. (Bohr 1935, p. 701)

Obviously, when an experimental arrangement is described in a classical way, it's described as being in a space of just three dimensions. I conclude that Bohr would not be on board with those who hold that quantum mechanics shows that the space that actually exists is $3N$-dimensional space. (For the record, I could provide a lot more evidence from Bohr to back this up.)

But Bohr was not the only developer of quantum mechanics; Schrödinger played a key role as well. Schrödinger does explicitly consider the possibility that the ontology for quantum mechanics involves a $3N$-dimensional space. In fact, one might think that he is endorsing that ontology when he writes: "The true mechanical process is realised or represented in a fitting way by the wave processes in q-space [where 'q-space' is Schrödinger's terminology for 'configuration space']" (Schrödinger 1926, p. 25). But he makes this claim in the context of a discussion of one-particle systems, where configuration space is just three-dimensional space. So what would he say about a multiparticle system? Schrödinger considers a two-particle system late in the same paper, but he offers only one sentence about the physical representation of the six-dimensional wave function: "The direct interpretation of this wave function of six variables in three-dimensional space meets, at any rate initially, with difficulties of an abstract nature" (Schrödinger 1926, p. 39). Schrödinger does not elaborate on what these difficulties are, but it's clear he is not endorsing the hypothesis that space is $3N$-dimensional.

Lorentz picks up on this problem with multiparticle systems. In 1926, he wrote a letter to Schrödinger, in which he says:

> If I had to choose now between your wave mechanics and the matrix mechanics, I would give the preference to the former, because of its greater intuitive clarity, so long as one only has to deal with the three coordinates x, y, z. If, however, there are more degrees of freedom, then I cannot interpret the waves and vibrations physically, and I must therefore decide in favor of matrix mechanics. (Lorentz in Przibram 1967, p. 44)

Schrödinger kept trying to develop an ontology for the wave function—there's a long and interesting story here, but to present it all would be outside the scope of this chapter. The short version of the story is that Schrödinger was looking for a way of having the wave function be a mathematical representation of physical processes in three-dimensional space. For example, he wrote a letter in response to Lorentz, and the first point he addresses is the issue of the multiparticle wave function. He writes: "I have been very sensitive to this difficulty for a long time but believe that I have now overcome it" (Schrödinger in Przibram 1967, p. 55). One way to overcome the difficulty would be to decide that it's not a difficulty at all and embrace the thought that physical reality really consists of a wave function evolving in 3N-dimensional space. This is definitely not what Schrödinger did. Instead, he gave a (somewhat complicated) proposal for how the wave function can be understood as providing a representation of processes in three-dimensional space. I don't completely understand the proposal, and Schrödinger ultimately decided it was unsuccessful, but here is the proposal he gave in the letter to Lorentz:

> $|\psi|^2$ (just as ψ itself) is a function of 3N variables or, as I want to say, of N three dimensional spaces, R_1, R_2, ..., R_N. Now first let R_1 be identified with the real space and integrate $|\psi|_2$ over R_2, ..., R_N; second, identify R_2 with the real space and integrate over R_1, R_3, ..., R_N; and so on. The N individual results are to be added after they have been multiplied by certain constants which characterize the particles, (their charges, according to the former theory). I consider the result to be the electric charge density in real space. (Schrödinger in Przibram 1967, pp. 55–56)

Schrödinger gives a partial ontology for the wave function, showing how electric charge density in three-dimensional space can be determined via the wave function. Though it would be interesting to explore in more detail how this proposal is meant to work, the key point for our purposes is that Schrödinger is looking for a way to understand the wave function as representing what's going on in "real," three-dimensional space.

Schrödinger kept working on this project for a while, but by 1935 he had given up. He wrote: "I am long past the stage where I thought that one can consider the ψ-function as somehow a direct description of reality" (Schrödinger in Fine 1996, p. 82). For the record, it is unclear to me to what extent he gave up on the project of considering the wave function as a direct description of reality because of the measurement problem, and to what extent he gave up on the project because of the issues of interpreting the 3N-dimensional wave function as representing something existing in real, three-dimensional space. Clearly, though, Schrödinger was not willing to endorse the view that the space of reality is 3N-dimensional.

3 Interpreting Quantum Mechanics

Let's step back. We started this discussion of Bohr and Schrödinger because we were asking about how many dimensions space has, according to the false theory of quantum mechanics. Just as we look to Aristotle to determine the content of Aristotelian physics, it seems reasonable to look to Bohr and Schrödinger to determine the content of quantum mechanics. If we do that, though, we can readily conclude that people like Albert are wrong to hold that quantum mechanics says that space is really $3N$-dimensional—that's not the view that Bohr and Schrödinger endorsed.

It's open to people like Albert, though, to hold that the originators of quantum mechanics are not the final arbiters of the content of quantum mechanics. This point could be made generally about scientific theories—like political constitutions, scientific theories are living documents, and how to understand them evolves as history progresses. Though I would not be inclined to endorse this view about scientific theories (or constitutions) in general, quantum mechanics is a special case. The reason is that quantum mechanics as originally formulated has been deemed unacceptable by many physicists and philosophers of physics—it faces the measurement problem, and the originators of quantum mechanics did not come up with any acceptable solution to that problem. As a result, new versions of quantum mechanics have been put on the table. Three prominent versions, which I focus on in turn later, are Bohm's theory, modal interpretations, and spontaneous localization theories like the GRW theory. But when we look to the originators of these versions of quantum mechanics to obtain guidance as to how to understand the ontologies of these versions, we again do not get support for the hypothesis that quantum mechanics should be understood as saying that space is $3N$-dimensional.

Let's start with Bohm. According to Bohm's theory, particles always have definite positions and evolve deterministically in accordance with a dynamic equation of motion that involves the wave function. The wave function is sometimes referred to as a "pilot wave," pushing the particles around. This understanding of the wave function ignores the fact that the wave function is defined over $3N$-dimensional space, while the Bohmian particles evolve in three-dimensional space. Bohm recognized this problem. In his 1957 book, he first presents his theory for one electron, where the wave function for the electron would evolve in three-dimensional space (since $N = 1$). He then writes:

> a serious problem confronts us when we extend the theory ... to the treatment of more than one electron. This difficulty arises in the circumstance that, for this case, Schrödinger's equation (and also Dirac's equation) do not describe a wave in ordinary three-dimensional space, but instead they describe a wave in an abstract $3N$-dimensional space, when N is the number of particles. While our theory can be extended formally in a logically consistent way by introducing the concept of a wave in a $3N$-dimensional space, it is evident that this procedure is not really acceptable in a physical theory, and should at least be regarded as an artifice that one uses provisionally until one obtains a better theory in which everything is expressed once more in ordinary three-dimensional space. (Bohm 1957, p. 117)

Bohm doesn't elaborate on why using 3*N*-dimensional space is not really acceptable in a physical theory, but I take it that his reasoning is that a physical theory is supposed to be about physical reality, and in our world physical reality consists of ordinary three-dimensional space.

As Jeffrey Bub (1997) spells out, Bohm's interpretation is just one version of a modal interpretation. Modal interpretations specify when and which properties of particles are definite—but unlike Bohm's theory, these definite properties could be properties other than position. The key point though of modal interpretations is that they specify the properties that particles have in three-dimensional space. I don't see how one could provide a version of modal interpretations that made sense in the context of 3*N*-dimensional space. Perhaps it could be done, but this certainly is not a project in which proponents of modal interpretations have engaged.

The GRW theory is more promising from the standpoint of a proponent of the 3*N*-dimensional space ontology. In the GRW theory, the wave function evolves according to a modified version of Schrödinger's equation, where sometimes the wave function indeterministically spontaneously collapses. If this is all there is to the ontology of the GRW theory, then it indeed endorses the hypothesis that the universe is really 3*N*-dimensional. But in fact, Ghirardi himself (the *G* of GRW, and the leading proponent of the theory) wants to add more to his theory, as he makes clear in for example Ghirardi, Grassi, and Benatti (1995). Ghirardi is often interpreted as endorsing the "accessible mass density" link, which specifies how the mass of objects in *three*-dimensional space is distributed, given the structure of the wave function. I have argued (Monton 2004) that this is not the best ontology for the GRW theory, and instead have endorsed the "mass density simpliciter" link, which specifies a somewhat different distribution of the mass of objects in three-dimensional space. The key point is that this debate is happening in the context of understanding what the GRW theory says about what's going on in three-dimensional space; the 3*N*-dimensional ontology is not being endorsed (at least not by Ghirardi).

4 The Wave Function Is Represented by a Property

People like Albert, who endorse the hypothesis that space is really 3*N*-dimensional, could just say that people like Bohr, Schrödinger, Bohm, and Ghirardi are wrong to understand quantum mechanics in the way that they do. But that sounds a bit like saying that Aristotle was wrong to understand Aristotelian physics in the way that he did. (Aristotle was wrong about the truths about physics, but he wasn't wrong about the content of Aristotelian physics.) Perhaps people like Albert should instead be viewed as presenting a new version of quantum mechanics, modifying whatever version they want to start with. For example, starting with Bohm's theory, they could argue that Bohm is wrong to hold that the concept of a wave in 3N-dimensional space is not really acceptable in a physical theory, and they can offer a new version of Bohm's theory with the 3N-dimensional ontology. Starting from the GRW theory, they can set aside the debate about the correct version of the mass density link and hold that in fact all that exists according to the GRW theory is the wave function evolving in 3N-dimensional space. People like Albert can then argue that their new version of quantum mechanics, with the 3N-dimensional ontology, is better than all the previous versions of quantum mechanics that have been proposed.

This brings us to the key problem with the 3N-dimensional ontology: there is no good reason to endorse it. All the work that the physically existing wave function does can instead be done by a property of the system of all the particles in three-dimensional space (as I first pointed out in Monton 2006). As I discuss in more detail in the next section, given the choice between a radically revisionary 3N-dimensional ontology and a normal three-dimensional ontology where the N particles in the universe collectively have a certain property, we have no good reason to endorse the radically revisionary ontology.

What is this property I am postulating? Is this some special property that's never been discussed in the literature before, that I'm just making up? On the contrary, this property exists according to a standard way of understanding quantum mechanics. Specifically, it's standard to interpret quantum mechanics in such a way that the *eigenstate-to-eigenvalue link* is true, and according to that link, the N-particle system has a property that carries all the information represented by the wave function.

Here's how this works. Consider the wave function for an N-particle system; I grant that the mathematical description of the wave function is as of a field evolving in 3N-dimensional space. But the wave function is a representation of the quantum state of that N-particle system. This quantum state is the eigenstate of some observable. Now, what the eigenstate-to-eigenvalue link holds is that if the N-particle system is in an eigenstate of some observable, then the N-particle system actually has the property corresponding to that eigenstate. The observable can take various possible values, and the idea is that the property the system has is that it has the value—the "eigenvalue"—corresponding to that eigenstate. In other words, we do not need the wave function as a physical field evolving in a physically existing 3N-dimensional space—all the information about the system that the wave function carries can be carried by a single property of the N-particle system in physically existing three-dimensional space.

Moreover, moving from the wave function to a property isn't some special move I made up—it's just a fact about quantum mechanics that the wave function is a representation of the quantum state, and a standard way of understanding the quantum state of a system is that it corresponds to a property that a system has. My ontology uses this standard (yet, in this context, unappreciated) way of thinking. To sum up: on my ontology, the wave function doesn't exist as a physical field in physically existing 3N-dimensional space; it is represented by property possessed by the physically existing N-particle system in physically existing three-dimensional space.

For a more precise formulation of my view, it helps to think of the theory of quantum mechanics using the semantic view of scientific theories. In the semantic view, a theory consists of two parts: a set of mathematical models, and a theoretical hypothesis that says how those mathematical models are taken to represent the world. There are different ways to mathematically model quantum mechanics. For example, using the Hilbert space representation, the state of a system is given by a vector in Hilbert space, whereas using the Schrödinger representation, the state of a system is given by a wave function in $3N$-dimensional space. How does this correspond to the world? One could put forth a theoretical hypothesis saying that Hilbert space is physically real, and there really is a line of a particular length pointed in a particular direction in that space. Similarly, one could put forth a theoretical hypothesis saying that $3N$-dimensional space is physically real, and there really is a wave function field evolving in that space. I do not endorse either of these theoretical hypotheses. The theoretical hypothesis I endorse says that there is an N-particle system evolving in three-dimensional space; this N-particle system has a certain property, and that property can be mathematically represented by a field in $3N$-dimensional space or by a vector in Hilbert space. What is physically real, though, is the property, not the field or the vector. (Those who are mathematical Platonists are welcome to believe that the field or the vector exists as an abstract object; I myself am a nominalist so will set Platonism aside.)

An analogy might help in this context. Consider the color observable, and consider an ordinary object that has a particular color property. This color property can be mathematically represented by a point in a multidimensional color space. But I do not believe that multidimensional color space exists; what I believe exists is the ordinary object and the color property. (The metaphysics of properties gets tricky here, but I have said all I need to say to present my perhaps helpful analogy.)

So let's go back to quantum mechanics—what role does this property that I'm attributing to N-particle systems play? The dynamical evolution of a system in quantum mechanics is given by Schrödinger's equation, and Schrödinger's equation uses the quantum state of a system. Where is this information about the quantum state represented in the world? According to proponents of the $3N$-dimensional ontology, this information is represented by the wave function field in $3N$-dimensional space. I maintain, in contrast, that this information is represented by this property that the N-particle system has. (Just as the wave function evolves through time, the property that the N-particle system has changes.)

In his original paper promulgating the $3N$-dimensional ontology, Albert (1996, p. 283) writes: "insofar as we are committed to *realism*, there was simply never anything other than physical objects that wave functions *could* have been."

I maintain that that's mistaken—we can be committed to realism but hold that the wave function is represented in reality not by a physical *object* but by a *property*. Specifically, the wave function corresponds to a property of the N-particle system, and all the information carried by the wave function is instantiated in reality by that property of the N-particle system. We do not need $3N$-dimensional space to be real to interpret quantum mechanics realistically.

A passage from J. S. Bell backs me up on this. Bell is sometimes presented as supporting the $3N$-dimensional ontology, but in this passage, at least, he supports my view regarding the wave function: "we can regard it simply as a convenient but inessential mathematical device for formulating correlations between experimental procedures and experimental results, i.e., between one set of beables and another" (Bell 1987, p. 53). The wave function, according to Bell, is an inessential mathematical device; the beables, existing in three-dimensional space, are what's real. What I make explicit is the physical way the information carried by the wave function is represented in the world—the information is represented by a property had by the system of beables in three-dimensional space.

5   Comparing Ontologies

We have two ontologies on the table—the three-dimensional ontology and the $3N$-dimensional ontology. Why should we favor one over the other? Well, suppose that these two ontologies make the same empirical predictions. That is, suppose that all the experiences we have will be the same regardless of which ontology is true. That is, suppose that the correct relationship between consciousness and the physical world is such that a wave function evolving in $3N$-dimensional space can give rise to normal conscious experience. (I take issue with these suppositions in Monton 2002, but for the purposes of this chapter I set those arguments aside.) How, then, can we adjudicate between the ontologies? One mode of adjudication is how a choice of ontology will influence the development of future theories; I talk about that in the next section. Another mode is which ontology better fits the pragmatic virtues that scientists use, such as simplicity, elegance, ease of use, and consilience with other theories. I discuss these pragmatic virtues here.

Let's start with simplicity, elegance, and ease of use. For all these pragmatic virtues, I think that the choice between the standard three-dimensional ontology and the $3N$-dimensional ontology is a draw. I don't see much difference in simplicity or elegance between a three-dimensional space with N particles and a $3N$-dimensional space with a wave function field. Postulating that the wave function is represented by a property of the N-dimensional system is not an ad hoc move, because the eigenstate-to-eigenvalue link is a commonly accepted part of quantum theory. Regarding ease of use, both ontologies are ontologies for quantum mechanics, so mathematically, Schrödinger's equation and the wave function can still be used to make predictions for measurement outcomes regardless of which ontology holds.

With regard to consilience with other theories, here I maintain that the three-dimensional ontology is a clear winner. Theories in chemistry and biology and other parts of physics all talk about the world as if it has three dimensions of space. General relativity, for example, provides models of manifolds that have three spatial dimensions and one time dimension.

Related to the pragmatic virtue of consilience with other theories is the pragmatic virtue of consilience with common sense. As I suggested elsewhere (Monton 2006), a pragmatic virtue that scientists use is that one should not accept theories that radically revise people's everyday understanding of the world when there are other, at least equally acceptable theories that do not entail such extreme revision. The 3N-dimensional ontology is radically revisionary: we think that the world around us has objects extended in exactly three spatial dimensions, but in fact there is no such three-dimensional space and no such three-dimensional objects. As Albert (1996, p. 277) writes, "whatever impression we have, say, of living in a three-dimensional space ... is somehow flatly illusory."

Some disagree with my claim that the 3N-dimensional ontology is radically revisionary. Wallace and Timpson write:

> While the wave-function realist will deny that 3-dimensional objects and spatial structures find a place in the fundamental ontology, this is not to say that the 3-dimensional objects surrounding us, with which we constantly interact, and which we perceive, think and talk about, do not exist, that there are not truths about them. It is just to maintain that they are emergent objects, rather than fundamental ones. But an emergent object is no less real for being emergent. (2010, pp. 705–6)

Despite Wallace and Timpson's confident assertion to the contrary, a wave function in 3N-dimensional space does not give rise to three-dimensional emergent objects. To argue for my view, it would help to see why Wallace and Timpson think otherwise. But it's not clear from their discussion what is meant to ground the claim that three-dimensional objects exist emergently. As I see it, they have two options.

The first way is to appeal to the fact that there are observers who *experience* three-dimensional objects, and given that that experience takes place, there are (emergently) three-dimensional objects. The second way rejects this and holds that emergence has nothing to do with experience. The second way holds that even in a 3N-dimensional universe with no experience at all, there could (emergently) exist three-dimensional objects. (Moreover, just to make clear, we are not talking about objects existing within a three-dimensional hypersurface of this higher-dimensional space; we are talking about a more complicated form of emergence, in which particular sets of three dimensions of the $3N$-dimensional space correspond to particular positions of particles in the emergent objects. For more on this point, see Ney 2010.)

If the claim of emergence is meant to be grounded in experience, then I offer the following argument by analogy against their position. Imagine that Wallace and Timpson try to appeal to emergence in the context of skepticism. Imagine that they maintain that we are brains in vats, and we don't have hands according to the fundamental ontology, but nevertheless we do have hands—hands are emergent objects. A view along these lines has been presented before, by David Chalmers (2005), but the vast majority of philosophers definitively reject this purported solution to skepticism. The reason this purported solution to skepticism should be rejected is that, given the brain-in-the-vat ontology, it's a fact about reality that the observers who are brains in vats don't have hands, despite the fact that they have the experience of having hands. The same claim can be made in the context of the $3N$-dimensional ontology—it's a fact about reality that there aren't three-dimensional objects, despite the fact that observers have the experience of interacting with three-dimensional objects.

Let's turn to the second way of understanding the claim that three-dimensional objects exist emergently. Maybe Wallace and Timpson hold that there's something special about the structure of the wave function in $3N$-dimensional space that gives rise to three-dimensional objects, even in a world in which there's no experience at all. I have two responses. First, I would need to see the argument. Second, I do not think one could provide a sound argument for this, because reality doesn't work that way. It's simply not the case that one can have a $3N$-dimensional space with a field evolving in it, such that when the field has a certain configuration, three-dimensional objects come into existence. Granted, this is not logically impossible—there could be laws of physics that specify the conditions under which three-dimensional objects come into existence—but for them to come into existence emergently, without this happening in accordance with certain novel laws of physics, is not the way a world where quantum mechanics is true works.

Similarly, for a three-dimensional Newtonian world with N point particles, it is unreasonable to hold that a single point particle in $3N$-dimensional space emergently exists, even though there is a sense in which the $3N$-dimensional configuration space with the single particle has a straightforward mathematical correspondence with the N point particles in three-dimensional space. (Also, given this three-dimensional Newtonian world with N point particles, it is unreasonable to hold that there emergently exist two point particles in a $3N/2$-dimensional space [assuming $3N$ is even], or four point particles in a $3N/4$-dimensional space [assuming $3N$ is a multiple of 4], and so on.)

So far I have been engaging in speculation—but what is Wallace and Timpson's actual argument for the claim that three-dimensional objects exist emergently, given the 3*N*-dimensional ontology? Unfortunately, they don't provide much of one. Just after the passage I already quoted, they continue with the following:

> It is also worth keeping in mind that many workers in quantum gravity have long taken seriously the possibility that our 4-dimensional spacetime will turn out to be emergent from some underlying reality that is either higher-dimensional (as in the case of string theory) or not spatio-temporal at all (as in the case of loop quantum gravity). In neither case is it suggested that ordinary spacetime is *non-existent*, just that it is *emergent*. (Wallace and Timpson 2010, p. 706)

In the string theory case, I believe that four-dimensional space-time exists, but I wouldn't say it's emergent. Instead, the three spatial dimensions we are familiar with exist as a kind of hypersurface in a higher-dimensional space— the other spatial dimensions are such that we don't perceive ourselves as moving through them. The loop quantum gravity case is different, because on that theory fundamental reality is not spatiotemporal at all. The passive voice construction of Wallace and Timpson's last sentence hides the fact that a theory like loop quantum gravity is open to philosophical interpretation and that some philosophers—I, for example—would strongly argue that for a nonspatiotemporal theory like loop quantum gravity, then indeed ordinary space-time is nonexistent.

Thus, I conclude that the 3*N*-dimensional ontology does not include the existence of real yet emergent three-dimensional objects. It follows that the 3*N*-dimensional ontology really does provide a radically revisionary account of the world, and this is a pragmatic mark against it.

6 Looking Ahead

I maintain that my wave-function-represented-by-a-property-of-an-*N*-particle-system ontology is better than the wave-function-field-evolving-in-3*N*-dimensional-space ontology. But I do not want to argue that my ontology is right and the other ontology is wrong—quantum mechanics is a false theory, so it is natural to conclude that any ontology for quantum mechanics is a false ontology.

Although these ontologies may be false in all their details, one may be more on the right track than the other. It may be that quantum mechanics is false, but we really are living in a space with a large number of dimensions, such that if we were presented with the true theory of physics we would see a natural connection between the quantum-mechanical ontology of 3*N*-dimensional space and the ontology of the true theory. I think that this is ultimately the viable and prima facie promising claim that the proponents of the 3*N*-dimensional ontology should be understood as making.

It's part of the history of physics that physicists will identify certain claims in a theory as being definitively true, even when they recognize that the theory itself is false. Some false theories are taken to provide certain insights that will carry over into the development of any future theories. For example, Copernican cosmology endorsed the view that Earth is not at the center of the universe, and that view is universally taken to be an insight of this false theory that carries over into any future development of physics. Somewhat more controversially, most physicists hold that the idea that simultaneity is relative is a core idea of the false theory of special relativity, an idea that will get carried over into any future development of physics. Similarly, the proponent of the $3N$-dimensional ontology could hold that the idea that the space in which things fundamentally exist is configuration space is an insight that will get carried over into any future development of physics.

It's too early to say whether this idea will catch on the way the antigeocentrism and relativity of simultaneity ideas did, but in principle that could happen. But even if the $3N$-dimensional idea doesn't catch on in that way, it could still be fruitful. Specifically, proponents of the $3N$-dimensional ontology can be taken to be providing an expansion of possibilities. Before the development of their view, we had not even recognized that this was a possibility for how to interpret a physical theory. Now that we do, this is a possibility that can be kept in mind as new physical theories are developed in the future. The debate we engage in regarding whether the $3N$-dimensional ontology is the best ontology for quantum mechanics can be construed as an implicit debate regarding how seriously this possibility should be kept in mind for future theories of physics.

References

Albert, David (1996), "Elementary Quantum Metaphysics," in James Cushing, Arthur Fine, and Sheldon Goldstein (eds.), *Bohmian Mechanics and Quantum Theory: An Appraisal*, Dordrecht: Kluwer.

Bell, J. S. (1987), *Speakable and Unspeakable in Quantum Mechanics*, Cambridge: Cambridge University Press.

Bohm, David (1957), *Causality and Chance in Modern Physics*, London: Routledge & Kegan Paul.

Bohr, Niels (1935), "Can Quantum-Mechanical Description of Physical Reality Be Considered Complete?," in J. A. Wheeler and W. H. Zurek (eds.), *Quantum Theory and Measurement*, Princeton, N. J.: Princeton University Press.

Bub, Jeffrey (1997), *Interpreting the Quantum World*, Cambridge: Cambridge University Press.

Chalmers, David (2005), "The Matrix as Metaphysics," in C. Grau (ed.), *Philosophers Explore the Matrix*, Oxford: Oxford University Press.

Fine, Arthur (1996), *The Shaky Game: Einstein, Realism, and the Quantum Theory*, Chicago: University of Chicago Press.

Ghirardi, G. C., Grassi, R., and Benatti, F. (1995), "Describing the Macroscopic World: Closing the Circle within the Dynamical Reduction Program," *Foundations of Physics* 25:313–28.

Hafele, J., and Keating, R. (1972a), "Around the World Atomic Clocks: Predicted Relativistic Time

Gains," *Science* 177(4044):166–68.

Hafele, J., and Keating, R. (1972b), "Around the World Atomic Clocks: Observed Relativistic Time Gains," *Science* 177(4044):168–70.

Monton, Bradley (2002), "Wave Function Ontology," *Synthese* 130:265–77.

Monton, Bradley (2004), "The Problem of Ontology for Spontaneous Collapse Theories," *Studies in History and Philosophy of Modern Physics* 35:407–21.

Monton, Bradley (2006), "Quantum Mechanics and 3N-Dimensional Space," *Philosophy of Science* 73:778–89.

Monton, Bradley (2010), "Prolegomena to Any Future Physics-Based Metaphysics," in Jonathan Kvanvig (ed.), *Oxford Studies in Philosophy of Religion Volume III*, Oxford: Oxford University Press.

Ney, Alyssa (2010), "The Status of Our Ordinary Three Dimensions in a Quantum Universe," *Noûs*, doi: 10.1111/j.1468–0068.2010.00797.x.

Przibram, K. (ed.) (1967), *Letters on Wave Mechanics*. New York: Philosophical Library.

Schrödinger, Erwin (1926), "Quantization as a Problem of Proper Values, Part II," in Erwin Schrödinger, *Collected Papers on Wave Mechanics*, New York: AMS Chelsea Publishing.

Wallace, David, and Timpson, Christopher (2010), "Quantum Mechanics on Spacetime I: Spacetime State Realism," *British Journal for the Philosophy of Science* 61:697–727.

8

Ontological Reduction and the Wave Function Ontology*

ALYSSA NEY

An increasingly influential position among philosophers of physics and metaphysicians is wave function realism: the view that the wave function of quantum mechanics is a real, fundamental entity. As metaphysicians begin to come to terms with what would be the consequences of this intriguing fundamental ontology, a lively debate has sprung up within the philosophy of physics about whether wave function realism is a plausible metaphysical interpretation of quantum mechanics (see, for instance, Albert 1996; Monton 2002, 2006; Maudlin 2007, in addition to the essays in the present volume). The goal of this chapter is not to argue either for or against wave function realism, but rather to investigate an important issue that must be settled prior to deciding between these rival fundamental ontologies for quantum mechanics. This issue concerns how one might capture in what sense, if any, familiar macroscopic objects (tables, chairs, people, mental states, and so on) may be reduced to an ontology in which one of the fundamental objects is the wave function interpreted realistically. Before we can ask whether wave function realism is motivated, we must ask whether there is any coherent way to understand how macroscopic objects (including ourselves) may exist in a world that is fundamentally made up of a wave function. In what follows, I explore the challenges traditional models of ontological reduction face in providing such an account. I then make a tentative proposal toward a new conception of ontological reduction that may better serve this purpose.

* I thank audiences at the University of Miami, University of Nevada at Las Vegas, the Southern Society for Philosophy and Psychology's 2011 meeting in New Orleans, and the 2011 Metaphysics of Science conference at the University of Toronto where this paper was presented. I especially thank Otavio Bueno, Greg Janssen, James Ladyman, Peter Lewis, Colin McGinn, Bill Ramsey, and James Woodbridge for helpful comments and discussion.

1 The Wave Function Ontology

Let's start by introducing the main features realists take the wave function to have. Although its name may appear to imply otherwise, wave function realists (for example Bell 1987, p. 128; Albert 1996; and Lewis 2004) deny that the wave function is an abstract, mathematical object. Instead, they typically claim that the best realist understanding of quantum mechanics takes the wave function to be a field, but unlike more familiar electromagnetic fields, it is spread out not over three-dimensional space or four-dimensional space-time, but over a much higher-dimensional space. Wave function realists have put this point this way:

> No one can understand this theory [Bohmian mechanics] until he is willing to think of Ψ [the wave function] as a real objective field.... Even though it propagates not in 3-space but in $3N$-space. (Bell 1987, p. 128)
>
> There is nothing in this theory [Ghirardi-Rimini-Weber, or GRW] but the wavefunction. It is in the wavefunction that we must find an image of the physical world, and in particular of the arrangement of things in ordinary three-dimensional space. But the wavefunction as a whole lives in a much bigger space, of $3N$ dimensions. (Bell 1987, p. 204)[1]

In addition,

> The sorts of physical objects that wave functions *are* ... are (plainly) *fields*—which is to say that they are the sorts of objects whose states one specifies by specifying the values of some set of numbers at every point in the space where they live, the sorts of objects whose states one specified (in *this* case) by specifying the values of *two* numbers (one of which is usually referred to as an *amplitude*, and the other as a *phase*) at every point in the universe's so-called *configuration* space. (Albert 1996, p. 278)

The wave function is a field in the sense that it is spread out completely over the space it inhabits, possessing values, amplitudes in particular, at each point in this space.

The high-dimensional space in which the wave function exists is what physicists refer to as "configuration space." Traditionally, "configuration space" refers to an abstract space that is used to represent possible configurations of particles in three-dimensional space. Each point in the configuration space represents a possible spatial configuration and may be labeled by an ordered n-tuple in which

---

[1] Both of these passages from Bell are also quoted in Lewis (2004, p. 718) in defense of this interpretation of quantum mechanics.

the first three members represent the three spatial coordinates of the first particle, the second three members represent the three spatial coordinates of the
second particle, and so on. For a collection of N particles then, the corresponding configuration space is $3N$-dimensional.[2] Each point in this $3N$-dimensional
space $(x_1, y_1, z_1, x_2, y_2, z_2, \ldots, x_n, y_n, z_n)$ represents a state in which the n particles
are arranged so as to be at locations (x_1, y_1, z_1), (x_2, y_2, z_2), $\ldots$, and (x_n, y_n, z_n).

For wave function realists, this is the fundamental space of the universe. It
is the space over which the wave function is spread. Its dimensionality may be
captured heuristically by the formula $d = 3N$, where N is the number of particles
in the world, but this is merely an heuristic. The proper way to understand the
dimensionality of configuration space is in terms of the number of degrees of
freedom needed to accurately capture the quantum state of the universe.

Sometimes it can seem as if the philosophical problems raised by quantum
mechanics are just ones of interpretation—that physicists are in possession of
an uncontentious set of equations and the only question is how to interpret or
understand them. However, one central point of disagreement in the philosophy
of quantum mechanics concerns which set of equations is the correct one for
physics to use. Versions of quantum mechanics fall into two kinds: no-collapse
versions (like Everett's "many worlds" interpretation) in which quantum systems always evolve deterministically according to Schrödinger's equation, and
collapse versions (like orthodox quantum mechanics or the GRW spontaneous
collapse theory) in which quantum systems sometimes collapse.[3] The difference between these versions is clearest when one considers scenarios in which
the Schrödinger equation predicts that systems will evolve into states that are
accurately described as superpositions of, for example, a particle going through
one slit in a screen or another or of a cat being alive and dead. No-collapse versions like Everett's will entail that the state of the system does not evolve out of
these kinds of states, so in some sense the system as a whole never becomes one
in which the particle determinately went through one slit or the other, or the
cat determinately was alive or dead.[4] On the other hand, versions of quantum
mechanics with a collapse postulate predict that systems will evolve out of such

---

[2] Unfortunately, for a view that takes the wave function to be an element of the fundamental
ontology, the label "configuration space" is misleading. According to wave function realism, particles in
three-dimensional space are not ontologically prior to the wave function, so the space the wave function inhabits is not fundamentally a space of configurations of particles in three-dimensional space.
Nevertheless, this terminology is entrenched, so I retain it here.

[3] I follow Albert and Loewer (1989) in drawing this distinction.

[4] Those familiar with the philosophy of quantum mechanics literature will notice that I do not
mention some widely discussed versions of quantum mechanics (e.g., Bohmian mechanics and modal
interpretations). I set these aside here because they contain equations describing more than just the
evolution of the wave function over time, so it is not clear that a wave function realist will face exactly
the same problems of reducing the world to the wave function understanding quantum mechanics
in these ways. This isn't to say the wave function realist won't face similar problems when trying to
understand these theories, I just do not consider those issues here.

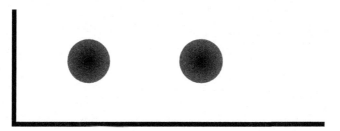

Figure 8.1

indeterminacy—the particle ends up determinately in one place or another and the cat evolves into a state in which it is either determinately alive or determinately dead. They say that the wave function collapses when a measurement or observation of a system takes place (as in the orthodox version of quantum mechanics by von Neumann 1932/1996), or (as in Ghirardi, Rimini, and Weber 1985) relate the probability of collapse at a given time to features of the quantum system itself, roughly its complexity.

To understand wave function realism, it may help to see some examples of what the world would be like fundamentally if this view were true. Consider Schrödinger's (1935) cat scenario. The world contains a cat sealed in a box along with a radioactive substance connected to a device set to release a poisonous gas if decay occurs. Let's assume that the half-life of the substance is such that quantum mechanics will predict a 50% chance of a decay occurring by noon. So there is a 50% chance of the poison releasing at noon and the cat dying, and a 50% chance of the poison failing to release and the cat being alive at noon.

A wave function realist disposed toward a no-collapse version of quantum mechanics will say that in situations like this of Schrödinger's cat, by noon, the wave function will evolve into a state with high amplitude in two regions of configuration space—one grounding the release of poison and a dead cat, and the other grounding no release of poison and a living cat. Dark regions in figure 8.1 are intended to represent regions of configuration space in which the amplitude of the wave function is high. *Each point in these regions of high-amplitude wave function corresponds to a particular classical configuration of particles. Each configuration is such that it would be sufficient to ground the existence of either a living cat or a dead cat, were the world a classical world. The large dot on the left in figure 8.1 represents all of the ways of there (classically) being a living cat in the box. The dot on the right in figure 8.1 represents all of the ways of there (classically) being a dead cat in the box.*[5] Because this is a quantum world, the state of the universe will not reduce to a wave function with high amplitude solely at one point in the configuration space—regions of high intensity of the wave function will instead be spread

---

[5] I emphasize these sentences because they will be drawn on below in the model of ontological reduction I propose.

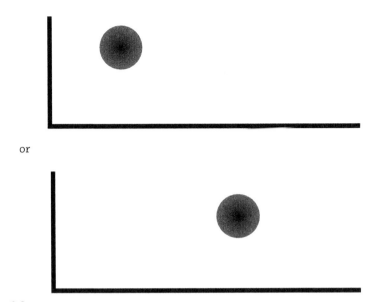

or

Figure 8.2

throughout this space (see Albert 1992, pp. 97–98). But it is possible to understand points in configuration space (in a top-down manner) as corresponding to particular ways the world might be if this were a classical world and all of the wave function's region of nonzero amplitude was instantiated at that point.

On the other hand, for a version of quantum mechanics with collapse, in Schrödinger's cat scenario, although the laws strictly speaking permit the wave function to evolve into a state like that we just described for Everettian quantum mechanics, it is extremely likely that the wave function will undergo a collapse. What this means is that although the wave function may for a time evolve toward a state with two separated regions of high amplitude, it is extremely likely that it will quickly collapse and evolve into a state with high amplitude in either one of these regions or the other. Either the wave function will be peaked around one region of configuration space at noon grounding a dead cat, or the wave function will be peaked around a distinct region at noon grounding a living cat (see figure 8.2). The fundamental ontology of the world then is a wave function spread out with one clumping of high amplitude in one region of configuration space.

In all, what this section is meant to illustrate is merely that for a wave function realist adopting Everettian quantum mechanics or a collapse version of quantum mechanics, the state of the world is constituted by the state of the wave function. What kind of objects there are, including what kind of macroscopic objects there are, will depend inter alia on facts about where the wave function possesses high amplitude. The next sections examine how one ought to understand this dependence.

2 Models of Ontological Reduction

Let's begin by assuming wave function realism. Then, if we want to be realists about macroscopic objects like tables, chairs, and people, we need to give an account of how these kinds of objects can exist in a world that is fundamentally made up of a wave function in configuration space.

What I am most interested in here is the project of facilitating an ontological reduction. When I talk about *ontological* reductions, the contrast is with the classical notion of inter-theoretic reduction in the philosophy of science. There, the primary interest is in showing how scientific theories may be unified. For example, one would like to know what is the relationship between special sciences like chemistry or psychology (the theories that are the targets of the reduction) and fundamental physics (the base theory). Here, as part of our post-positivist metaphysics of science program, we are primarily interested in how ontologies may be unified. In particular, the main issue is how macroscopic material objects (the ontology that is the target of the reduction) could be nothing over and above the wave function in configuration space (our base ontology).

It will be useful for what follows to note a distinction between kinds of ontological reduction—namely, that some ontological reductions are *eliminative* reductions (where the success of a reduction gives one a reason to eliminate the target class of entities) and others are *retentive* reductions (in which it is reasonable to retain the target class of entities in one's ontology after the reduction occurs). As several philosophers have emphasized,[6] real cases of ontological reduction actually instantiate a spectrum of cases ranging from the clearly eliminative on one end (e.g., the reduction of phlogiston to oxygen and other elements of the ontology of modern chemistry) to the clearly retentive on the other end (e.g., the reduction of lightning to electrical discharge in the atmosphere). Many cases of ontological reductions fall somewhere in between the two extremes, where in the process of trying to accomplish an ontological reduction one ends up having to revise some of what one had earlier believed about the nature of the target class of entities. We are used to expecting that our reduction of material objects to fundamental physical phenomena will lie somewhere in the middle of the spectrum (though hopefully closer to the retentive end). Consider the revision in conceptions of the solidity of material objects or the secondary qualities. After successful reductions of the past, most of us have long given up the view that colors are intrinsic features spread over the surfaces of objects.[7] Any successful reduction of phenomenal consciousness to a physical ontology will also likely lie somewhere toward the middle of the spectrum.

---

[6] See in particular the very nice chart in Bickle 1998, p. 30.

[7] We believe that our world is not David Chalmers's version of Eden, in which objects possess the features naive perception seems to represent them as having (Chalmers 2006).

In the course of a physicalist reduction, we will be forced to revise some of our previous beliefs about the nature of conscious states.

The first thing one might try in attempting a reduction of macroscopic objects to a wave function ontology is to look to those mainstream philosophical models of ontological reduction with which we are familiar and see if they can be of any use in this case. One long-popular picture comes from Paul Oppenheim and Hilary Putnam's "Unity of Science as a Working Hypothesis" (1958). These authors summarize their account of how one reduces a target ontology (that associated with a branch of a theory B_2) to a more fundamental, base ontology (that associated with a branch of a theory B_1) in the following way to yield an account of what they call "micro-reduction":

> Under the following conditions we shall say that reduction of B_2 to B_1 is a *micro-reduction*: B_2 is reduced to B_1; and the objects in the universe of discourse of B_2 are wholes which possess a decomposition ... into proper parts all of which belong to the universe of discourse of B_1. (Oppenheim and Putnam 1958, p. 6)

This account of micro-reduction states two requirements. First, one branch of science or theory is reduced to another. Oppenheim and Putnam conceive of branch reduction along the lines of earlier work of Kemeny and Oppenheim (1956).[8] To put their account of branch reduction roughly, Kemeny and Oppenheim think that a branch B_2 is reduced to a branch B_1 if the elements of the ontology of B_1 suffice to explain all of the observations that can be explained by appeal to the elements of the ontology of B_2. For example, chemistry has been branch-reduced to physics just in case all of the observations that one could explain by appeal to the entities of chemistry (molecules, liquids, gases, etc.) are also explainable by appealing instead only to the entities of physics (electrons, protons, electromagnetic fields, etc.).

The success of this branch reduction by itself does not give us reason to think the objects of chemistry are nothing over and above the entities of physics. This is why they need the second part of their account, because what B_1 and B_2 might produce are two wholly distinct explanations of our observations. Oppenheim and Putnam add an additional, distinctly metaphysical requirement for micro-reduction: that the ontology of the target theory possess a decomposition into elements of the ontology of the base theory. To continue our example, if we are

---

[8] Oddly, it is often assumed that Oppenheim and Putnam presupposed Ernest Nagel's (1961) theory of inter-theoretic reduction using bridge laws in their paper. My suspicion is that this confusion stems from Fodor's (1974) critique of Oppenheim and Putnam. In any case, Oppenheim and Putnam are explicit in their paper that they are assuming the Kemeny/Oppenheim model of inter-theoretic reduction described shortly (1958, p. 5). Kemeny and Oppenheim viewed Nagel's account of reduction as insufficiently narrow, not recognizing cases in which neat correspondences between terms in the target and base theories are impossible to obtain.

told that the chemical entities possess a decomposition into physical particles, we see how there is really only one explanation at work and that the entities of chemistry are nothing over and above the entities of physics. Moreover, this second mereological constraint in Oppenheim and Putnam's model ensures that micro-reductions be at least somewhat retentive as opposed to purely elimi-native: when one appeals to complex arrangements of physical particles and electromagnetism in an explanation of, for example combustion, the chemical phenomena are not thereby reasonably eliminated, because they are seen to just be the mereological composites of the microphysical entities in question.

One might object that only the mereological claim is needed to establish that material objects are nothing over and above subatomic particles, that is, to accomplish the ontological reduction. If this is the case, this will not affect my aims in the rest of the chapter. However, one might think that here the claim of branch reduction in the Kemeny/Oppenheim sense is also necessary to the reduction. For one might want to leave it as an open possibility that a whole may be something over and above its proper parts because the whole can explain phenomena the proper parts alone cannot. When we give the branch reduction as well, we establish that this is not the case, because any explanation in terms of the whole is ultimately dispensable in terms of explanation appealing only to the (proper) parts. My own view is that both parts of the account—the branch reduction and the metaphysical part—are required to accomplish the ontological reduction.

If Oppenheim and Putnam's constraints are satisfied, there is reason to believe one has accomplished a retentive ontological reduction.[9] However, unfortunately, when one comes to take the wave function ontology seriously, this model begins to look unhelpful. We cannot micro-reduce tables, cats, and other macroscopic objects to the wave function in the way described by Oppenheim and Putnam because in this case the sole element of our base ontology, the wave function, is not a proper part of a table or a cat. There may be a spatial decomposition of the wave function in configuration space. But there are not separate parts of the wave function corresponding to (say) the leg of a table or the head of a cat. Recall, even if we may talk about the left and right half of a region of high-amplitude wave function as in the figures in section 1, these cannot be viewed

---

[9] As has been well noted in the reduction literature, there are worries about the use of the Kemeny/Oppenheim model of branch reduction stemming from its positivist assumption of a sharp distinction between theory and observation. If one is concerned about this, one may substitute in one's own preferred account of branch reduction. One option would be to substitute the first two stages of Jaegwon Kim's (1998, pp. 97–103) or David Lewis's (1972) models of functional reduction. One first undertakes a functional analysis of the target ontology, characterizing it in terms of a network of causes and effects. Then one finds entities in the base ontology that can play these causal roles. The difference is that whereas in the Kemeny/Oppenheim model of branch reduction we are trying to capture directly the explanations of observations the target ontology allowed us, here we are trying to capture the constitutive causal features of the target ontology.

as left and right halves of a cat. Rather, each will only be a large collection of different ways the entire system could be, were the world classical. Thus, even if the first constraint for Oppenheim and Putnam micro-reduction may be satisfied, and we can explain everything we previously explained using the ontology of tables and cats by appealing to the elements of a wave function ontology, the second constraint will fail to be satisfied.[10] There is no mereological decomposition of ordinary macroscopic objects into the elements of a wave function ontology.

David Wallace is one philosopher who has been careful to note the limits of traditional models of reduction for the task of understanding the relationship between macroscopic objects and the ontology of quantum mechanics.[11] His response is to recommend a different model of reduction suggested by Daniel Dennett:

> This makes the pattern-based view of ontology espoused by Dennett (1991) ... very attractive. Of course, there must be some sense in which macroscopic objects are built out of microscopic constituents and in which they are supervenient on the properties of the constituents. Dennett, by regarding macro-objects as *patterns in* the micro-ontology, rather than as *mereological sums of* that micro-ontology, provides the sort of account of compositionality that is not hostage to contentious or downright false pictures of physics. (Wallace 2004, p. 635)

Dennett's model involves understanding the target entities in a reduction as patterns in the base domain. What is supposed to be appealing about this view (what attracts Wallace to it) is that it makes it relatively easy to give an account of the relation between elements of the microscopic and macroscopic ontologies that doesn't make assumptions about what the fundamental ontology looks like—assumptions that can be refuted by contemporary physical theories. As Wallace says, "This I find dangerous: It bets our metaphysical structure on the current state of fundamental physics, despite the fact that fundamental physics frequently changes" (2004, p. 635). Because vastly different kinds of

---

[10] I put things this way, but the situation is not quite so extreme as this. It is not clearly the case that the wave function realist must throw out the old Oppenheim and Putnam model of reduction entirely. Reductions may proceed in stages. (To reverse a metaphor of Bickle via Kenneth Schaffner, reductions may be creeping rather than sweeping.) We may first ask not how could a wave function undulating in configuration space ground the existence of objects like tables and chairs, but instead how could a wave function undulating in configuration space ground the existence of particles moving around in three-dimensional space. Then we can retain the Oppenheim and Putnam model of micro-reduction for getting us from the particles in three-dimensional space to tables and chairs.

[11] It may go without saying that appealing to supervenience instead of the part/whole relation in a model of reduction also fails to be illuminating here. Claiming that our manifest image supervenes on the behavior of the wave function does not settle questions of nothing over and aboveness (see Wilson 2005).

phenomena (pixels, stock market behavior, the weather) can instantiate patterns, the Dennett model doesn't presuppose that the base ontology contain localized, separable parts that can add up to macroscopic wholes by a relation of mereological composition.

Wallace takes the neutrality of Dennett's model to be a virtue; however, I argue, this model of reduction is so bleached out as to be unilluminating. Part of the task of an account of reduction is to provide a story of how it is the case that these objects we believe in, like tables and chairs and cats, could possibly be nothing over and above something as novel to our conceptual schemes and strange as whatever is posited by fundamental physics, in this case, the wave function. But this appeal to Dennett's account fails to specify the specific metaphysical relation grounding the putative fact of nothing-over-and-above-ness. There are some cases in which it is plausible to say that a class of entities has been shown to be nothing but patterns in some other class of entities. A clear case is Conway's Game of Life that Dennett originally used to motivate the real patterns account. But such applications of the pattern account depend on facts of mereological decomposition. In the Game of Life, the reduced entities (the gliders, blinkers, and so on) possess a clear mereological decomposition into the pixels that make up the reduction's base ontology. The question is whether a reduction is similarly achieved via this model when the patterns do not emerge from a spatial arrangement of the base ontology viewed as parts of which the patterns are wholes.

Moreover, what after all is so wrong with producing a model of reduction that is tied to the kind of fundamental physical ontology we have good reason to believe today? Sure, if we do so, we can count on our models of reduction having to be revised or supplemented in light of future drastic revisions to fundamental physics. But, so what? Drastic revisions to fundamental physics are always difficult to understand, and this seems to be one important part of the project of coming to understand them—seeing how familiar macroscopic objects could be nothing over and above the fundamental ontologies motivated by these theories.

3 Skepticism

Some, like Bradley Monton, have argued that the reductions I have in mind are hopeless and so we should reject wave function realism. One of his complaints relies on the fact that macroscopic objects like tables and cats occupy three-dimensional space—they have dimensions of height, width, and depth—but the wave function lives in an entirely separate $3N$-dimensional space in which there is no height, width, or depth (Monton 2006, pp. 783–84; also see his chapter in this volume). Monton's worry is that if wave function realism is true, then three-dimensional space is an illusion. If three-dimensional space is an illusion, then there cannot be any three-dimensional objects. So these objects too must be illusory:

> I conclude that the 3N-dimensional ontology does not include the existence of real yet emergent three-dimensional objects. It follows that the 3N-dimensional ontology really does provide a radically revisionary account of the world, and this is a pragmatic mark against it. (Monton's chapter in this volume)

This objection raises two questions for our project. (1) Is it true that if wave function realism is true, then the three-dimensional space of our ordinary experience must fail to exist? (2) Could macroscopic objects exist even if it were strictly speaking true that they do not have the three dimensions we thought they had? Although he does not distinguish these questions, Monton seems to take the answer to both of these questions to be "no."

However, neither of these questions have obvious answers. When it comes to the first question, several philosophers inclined toward wave function realism have argued that three-dimensional space is plausibly recoverable from the wave function ontology. Lewis (2004; see also his chapter in this volume) argues that configuration space is, in one sense at least, three-dimensional, even if it is 3N-dimensional in another sense. Albert has suggested that the existence of three-dimensional space may itself be retentively reduced to the diachronic behavior of the wave function by way of a functional reduction of our ordinary three spatial dimensions.[12]

However, to turn to the second question, it is worth noting that even if our ordinary three spatial dimensions cannot survive a reduction to a wave function ontology,[13] this still leaves open the possibility that macroscopic objects might. For even though these objects appear to have heights, widths, and depths, these features may turn out to be negotiable. Objects may still have enough of the features we associated with them after the reduction to a wave function ontology, even if they are no longer three-dimensional. In other words, this is to allow that the reduction of macroscopic objects to a wave function ontology may not be perfectly retentive if such an ontology does not suffice to ground the existence of three-dimensional space. It is arguable, the more one thinks about it, that no scientifically motivated ontological reduction is perfectly retentive, because

---

[12] In earlier work, he expressed a more eliminativist position about three-dimensional space, taking it to be an illusion (Albert 1996).

[13] This is a live possibility if substantivalism is true, because then spaces in general would appear to be fundamental entities and not the type of thing to be reduced to something else (e.g., configuration space, or the behavior of a wave function in configuration space as Albert 1996 envisions). I argue for this claim in Ney (2012).

reductions always involve revising some of what we had earlier thought about the reduced objects.[14]

In his most recent work, Monton makes a related point about some recent work of wave function realists devoted to the task of explaining how macroscopic objects might arise within such a fundamental ontology.[15] Monton focuses his discussion especially on the work of Albert (1996), who offers a story about how the behavior of the wave function over time might produce the appearance in us of a three-dimensional world. Albert claims that any physics that is going to have a chance at describing the world as we experience it is going to have to posit a wave function that evolves in such a way that it is able to play the functional role of a universe with three-dimensional particles in it (1996, pp. 279–80). Once we are able to ground our experience of three-dimensional particles, we can move from there to a reduction of macroscopic objects.

I will not reproduce Albert's account in detail here (but see 1996, pp. 280–81 as well as his contribution to this volume). The crux of the story is that to support our quasi-Newtonian manifest image, it must be the case that when the amplitude of the wavefunction gets high at certain locations in configuration space, the peaks begin to accelerate more quickly than they do at other kinds of locations. These locations correspond to situations in a classical world in which material objects come into contact. For example, in a nine-dimensional configuration space, when the wave function moves into a state with high amplitude around the location (1, 2, 3, 1, 2, 3, 1, 2, 3), for example, we should expect the peaks to accelerate rapidly through configuration space. This makes sense because this corresponds roughly in the classical three-dimensional representation to a system of three particles coming to all occupy location (1, 2, 3), that is, particles that are moving close enough to impact each other. We would expect this to create an acceleration in the movement of the particles in three-dimensional space. This corresponds to an expectation that the regions of high-amplitude wave function should accelerate in configuration space when they come close to locations like (1, 2, 3, 1, 2, 3, 1, 2, 3).

Monton raises a worry about Albert's proposal by appealing to the brain-in-a-vat scenario familiar from discussions of skepticism in epistemology and metaphysical anti-realism. In such a scenario, one assumes a system (a brain in a vat hooked up to a computer) that is able to explain all of the experiences one has of a world outside of one's mind without any of the objects in this external world actually existing (except perhaps one's brain). Monton argues that if it were sufficient for the wave function realist to countenance three-dimensional objects to just explain why we have the *appearance* of three-dimensional objects,

---

[14] I thank Bill Ramsey here.

[15] Monton intends this point as a continuation of his objection to wave function realism; however, I want to use his insight to make it clearer what the wave function realist needs to accomplish to reduce macroscopic objects to a wave function ontology.

we would have reason to believe (by analogy) that in a brain-in-a-vat scenario, one has hands. But of course even though the brain-in-a-vat scenario is intended to be a case in which we can understand how the ontology in question (brain, vat, computer, etc.) could produce the appearance of hands, this is not sufficient to make it the case that in such a scenario one has hands. So, Monton argues, even if Albert is successful in showing how the existence of a wave function ontology could produce the appearance of macroscopic objects, this isn't sufficient to make it the case that in a wave function ontology, there actually are any such things.

Monton makes an important point here. To tell a story about how the behavior of the wave function may produce the appearance of tables and chairs is not in principle to answer the question of how there could be tables and chairs on a wave function ontology. This isn't to say that such stories will be of no use to the wave function realist. There is also an issue about how we could be deceived into thinking the world is three-dimensional (if this is indeed an illusion). But this is where the project of this chapter comes in with the emphasis on the second part of Oppenheim and Putnam's analysis. What the wave function realist needs is a genuine, retentive reduction of the objects of our manifest image, not just an explanation for the appearances.

4 Grounding the World in the Wave Function

I propose to begin this project here by revising the traditional account of ontological reduction in light of the physical ontology in which we are interested. The wave function, we have said, is a field spread out on configuration space. Somehow the wave function grounds the existence of, for example, my desk. This is not because my desk is spread out all over configuration space with parts corresponding here to a leg and there to a top. Rather, the persistence of peaks of the wave function grounds the existence of my desk. Point-sized regions of these peaks correspond to slightly different (classical) ways of there being a desk there, slightly different configurations of particles that could make up a desk, among other things. If instead, the peaks of the wave function were centered in a disjoint region of configuration space, then my desk would not have existed, or there would have been a very different kind of desk.

On this account, macroscopic objects reduce to peaks in the wave function that cluster around certain regions of configuration space. We may thus propose a new model of ontological reduction that adjusts the Oppenheim/Putnam model in light of the kind of microscopic ontology we find in quantum mechanics:

> Under the following conditions we shall say that reduction of B_2 to B_1 is a *micro-reduction*: B_2 is reduced to B_1; and the objects in the universe of

discourse of B_2 permit a decomposition into precise states corresponding to classical configurations of particles each of which are instantiated by the object in the universe of discourse of B_1 to a sufficient
amplitude.

The account is based not on a mereological relation but on some distinct kind of
reductive, metaphysical relation. Here we are considering an object's decomposition not into its intuitive parts but into various modes at each point in configuration space that are instantiated to varying degrees. Each mode corresponds
to a slightly different classical version of itself. For example, my desk exists in
the wave function ontology in virtue of having many of these different modes
instantiated to a sufficient degree (amplitude) in the wave function.

Of course, this is only a start at an account of ontological reduction that can
be used by the wave function realist. It is an open question at this point whether
this is a model of ontological reduction that those who are skeptical of wave
function realism, like Monton, will accept. However, it goes some way toward
capturing accurately the metaphysical relationship between macroscopic objects
and the wave function of quantum mechanics, so it is a place to begin.

5 Closing Explanatory Gaps

I have now proposed a model of ontological reduction more suitable to capturing
macroscopic objects in the wave function ontology than the classic Oppenheim/
Putnam model of micro-reduction. One might look at the account, however, and
still wonder whether it could be sufficient to accomplish an ontological reduction. For one might be told (indeed, even come to believe) that macroscopic
objects bear the proposed metaphysical relation to the wave function and yet
still be puzzled. Even if we are told these peaks of high-amplitude wave function
ground the dead cat in a box, this may be hard to believe. There appears to be an
explanatory gap in the sense of Levine (1983). To paraphrase John Hawthorne,
who raises the worry for wave function realist interpretations of quantum mechanics in this form: "No matter how much we look back and forth at the [macro-
scopic] and [the wave function], we're just not going to be able to see why the
true principles are the true principles and why certain false principles are false"
(Hawthorne 2010, p. 148).

Addressing this worry will (I hope) help us see what is doing what in the proposed account. True, if we just "look" at the target ontology (macroscopic objects)
and the base ontology (the wave function) on their own, we will not see a connection between the two. Why should we take a wave function to genuinely ground a
cat in a box? However, the same could be said for what we thought were the correct physical micro-reductions before we considered quantum mechanics. If we
just "looked" at the target ontology (macroscopic objects) and that base ontology

(some particles), there would have been the same explanatory gap. Here we are considering the ground-level objects bare on their own without any of the nomic features or causal powers that let them behave the way they need to in order to ground macroscopic objects.

What closes the explanatory gap is filling in the first part of these accounts of micro-reduction. This is where we see how an appeal to a wave function in configuration space (or, if this were a true interpretation of physics, some particles floating around at great distances from one another) can explain all of the observations appeal to the macroscopic allows us. How the first part of the account of micro-reduction gets filled in will, of course, be complicated and involve a lot more physics. This will be necessary to close the explanatory gap between the microphysical and macroscopic domains.

References

Albert, David Z. 1992. *Quantum Mechanics and Experience*. Cambridge, Mass.: Harvard University Press.

Albert, David Z. 1996. Elementary Quantum Metaphysics. In *Bohmian Mechanics and Quantum Theory: An Appraisal*. J. T. Cushing, A. Fine, and S. Goldstein, eds. Dordrecht: Kluwer, 277–84.

Albert, David, and Barry Loewer. 1989. Two No-Collapse Interpretations of Quantum Theory. *Noûs* 23:169–86.

Bell, John. 1987. *Speakable and Unspeakable in Quantum Mechanics*. Cambridge: Cambridge University Press.

Bickle, John. 1998. *Psychoneural Reduction: The New Wave*. Cambridge, Mass.: MIT Press.

Chalmers, David. 1996. Perception and the Fall from Eden. In *Perceptual Experience*. T. Gendler and J. Hawthorne, eds. Oxford: Oxford University Press.

Dennett, Daniel. 1991. Real Patterns. *Journal of Philosophy* 88:27–51.

Fodor, Jerry. 1974. Special Sciences (or: The Disunity of Science as a Working Hypothesis). *Synthese* 28:97–115.

Ghirardi, Giancarlo, Alberto Rimini, and Tullio Weber. 1985. A Model for a Unified Quantum Description of Macroscopic and Microscopic Systems. *Physical Review* 34:470–91.

Hawthorne, John. 2010. A Metaphysician Looks at the Everett Interpretation. In *Many Worlds?* S. Saunders, J. Barrett, A. Kent, and D. Wallace, eds. Oxford: Oxford University Press.

Kemeny, John, and Paul Oppenheim. 1956. On Reduction. *Philosophical Studies* 7:6–19.

Kim, Jaegwon. 1998. *Mind in a Physical World*. Cambridge, Mass.: MIT Press.

Levine, Joseph. 1983. Materialism and Qualia: The Explanatory Gap. *Pacific Philosophical Quarterly* 64:354–61.

Lewis, David. 1972. Psychophysical and Theoretical Identifications. *Australasian Journal of Philosophy* 50:249–58.

Lewis, Peter. 2004. Life in Configuration Space. *British Journal for the Philosophy of Science* 55:713–29.

Maudlin, Tim. 2007. Completeness, Supervenience, and Ontology. *Journal of Physics A: Mathematical and Theoretical* 40:3151–71.

Monton, Bradley. 2002. Wave Function Ontology. *Synthese* 130:265–77.

Monton, Bradley. 2006. Quantum Mechanics and 3N-Dimensional Space. *Philosophy of Science* 73:778–89.

Nagel, Ernest. 1961. *The Structure of Science*. New York: Harcourt, Brace, World.

Ney, Alyssa. 2012. The Status of Our Ordinary Three Dimensions in a Quantum Universe. *Noûs* 46:525–560.

Oppenheim, Paul, and Hilary Putnam. 1958. Unity of Science as a Working Hypothesis. *Minnesota Studies in the Philosophy of Science* 2:3–36.

Schrödinger, Erwin. 1935. The Present Situation in Quantum Mechanics. Reprinted in *Quantum Theory and Measurement*. J. A. Wheeler and W. H. Zurek, eds. Princeton, N.J.: Princeton University Press, 1983.

Von Neumann, John. 1932/1996. *The Mathematical Foundations of Quantum Mechanics*. Princeton, N.J.: Princeton University Press.

Wallace, David. 2004. Protecting Cognitive Science from Quantum Theory. *Behavioral and Brain Sciences* 27:636–37.

Wilson, Jessica. 2005. Supervenience-Based Formulations of Physicalism. *Noûs* 39:426–59.

9

The Structure of a Quantum World*

JILL NORTH

What is a world governed by quantum mechanics fundamentally like? In particular, what is the fundamental *space* of such a world like?

This question is puzzling. For the wave function—the thing that's governed by the dynamical laws, the object whose evolution predicts the results of quantum mechanical experiments—occupies a space that is very different from the one we seem to live in. The wave function's space has many dimensions—many more than the three dimensions of ordinary space (or four dimensions of ordinary space-time[1]). Prima facie, realism about quantum mechanics seems to require realism about the wave function and the space it inhabits. But then there's a glaring question as to why, if our world is fundamentally quantum mechanical, we seem to live in a space of only three dimensions.

I argue for a view related to what has come to be called "wave function realism," though I wish to put the emphasis in a different place. Rather than starting from questions about the ontological status of the wave function (as do other authors who arrive at a view similar to the one I defend[2]), I want to focus on the fundamental space of a quantum mechanical world. Wave function realism will naturally go along with the view, but I will first argue for realism about the space on which the wave function lives. The reason is that there are some very general principles, familiar from elsewhere in physics, supporting the view that this space exists and is fundamental to such a world—that this is the fundamental *physical* space of such a world.

The question about the fundamental space of a quantum world is complicated in two ways that I leave aside here. First, the fundamental structure of a world's

* I am grateful to David Albert, Otávio Bueno, Alyssa Ney, Ted Sider, Christian Wüthrich, audience members at the Pacific APA in 2010 and Calvin College in 2012, and the Yale philosophy department for helpful discussion and comments on earlier versions of this essay.

[1] I drop the qualification from now on; it should be understood. Similar questions arise for four-dimensional space-time as for three-dimensional space.

[2] See especially Albert 1996.

space(time) may be more properly given by a relativistic theory. Still, it is plausible that the fundamental theory of our world will be quantum mechanical. So it is worthwhile to think about what the world's fundamental space would be if ordinary quantum mechanics is its fundamental theory, leaving aside relativistic quantum mechanics or some other final theory. More important, similar considerations should carry over to the relativistic case; for ease of discussion, I limit this chapter to ordinary quantum mechanics of particles.[3]

Second, there are different theories of quantum mechanics on the table. This will not affect the thrust of the discussion here, however. For all theories make central use of the wave function, and this suffices to generate the questions about space. Different theories disagree on whether there are other things in the world besides the wave function; and there is room for debate, on any theory, about what sort of thing the wave function is. Yet the discussion here takes off from the nature of the space that all these theories need to define the wave function.[4]

This chapter goes as follows. I first discuss the guiding principles I rely on, and the ways we use them in our scientific theorizing (section 1). I then argue that these principles support the conclusion that the wave function's space is fundamental to a quantum world (sections 2 and 3). I end by suggesting that there is a way of reconciling the fundamentality of this high-dimensional space with the three-dimensionality of our experience (section 4). Note that for the purpose of this discussion, I assume realism about quantum mechanics, so that the wave function directly represents or governs (at least part of; see note 4) the ontology of a quantum mechanical world; I will not be arguing for this here.

---

[3] Wallace and Timpson (2010) argue that the case for configuration space realism weakens in quantum field theory because (among other reasons) particles are not fundamental and particle positions are imprecise, so the very idea of a configuration space is unclear. On my view, however, the fundamental space is the wave function's space, not configuration space; see section 2.

[4] All theories, realistically construed, regard the wave function as directly representing or governing some part of the fundamental ontology. Some theories, though, posit more in the fundamental ontology than what's represented in the wave function; Bohm's theory also posits particles (or one "world particle"; more later). And on some understandings of Bohm's theory—and some understandings of collapse theories like Ghirardi-Rimini-Weber (GRW) (Allori et al. 2008)—the theory only posits particles in the physical ontology; the wave function then governs the particles. Different theories also disagree on the dynamics of the wave function. They all contain the Schrödinger equation as a fundamental law, but some theories also have an indeterministic law of wave function collapse, as in GRW. Bohm's theory contains an additional dynamical law, the guidance equation, though this can arguably be derived from the Schrödinger equation plus symmetry considerations. Many-worlds theories add nothing to the fundamental dynamics or ontology beyond the standard formalism and the wave function. On any theory, it is open to debate whether the wave function alone represents the fundamental ontology, or whether there are also, or instead, objects in ordinary space(time). There are different ways of construing the wave function—as a field, a law, a global property of particles, as belonging to another metaphysical category—but the wave function is invariably central.

I have a sneaking suspicion that there will ultimately be a stand-off between ordinary space and wave function space views, depending on what sort of evidence one chooses to weight most heavily: whether from ordinary experience or our usual inferences from fundamental physics (see the end of section 3). I suspect there is no conclusive argument that one type of evidence or the other must be primary—no conclusive argument for whether to privilege the manifest or scientific image of the world when these come into conflict. I present an argument intended to resolve the stand-off in favor of the scientific image, but I doubt this should convince the opponent who starts off prioritizing the manifest one. My goals here are more modest: to present a case for wave function space realism and to defray the most counterintuitive consequences of the view.

1 The Dynamics as Guide to What Is Fundamental

The principles I will use to argue for the fundamentality of the wave function's space can be summed up in slogan form like this: the dynamical laws are a guide to the fundamental nature of a world.

Spelling this out in more detail: how is a world built up, according to its fundamental physics?

At the fundamental level, there is the fundamental ontology of the theory, there is the space in which this ontology lives, and there is some structure to that space. Then there are the dynamical laws, which say how the ontology evolves through this space over time.

This brings me to a very general principle that guides our physical theorizing, from which the other principles I use all extend: the dynamical laws are about what's fundamental to a world. The dynamical laws relate what's fundamental to what's fundamental, where what's fundamental includes the fundamental space and its structure, and the fundamental ontology. The dynamical laws govern the fundamental level of reality; that is why they are a guide to the fundamental nature of a world.

When I say that the laws "are a guide to the fundamental nature of a world," I mean that we infer the fundamental nature of a world from the dynamical laws. We do not directly observe the fundamental level of reality: we infer it from the dynamics. We posit, at the fundamental level, whatever the dynamical laws presuppose—whatever there must be in the world for these laws to be true of it.[5]

---

[5] Compare Albert 1996 on the dimensionality of ordinary space. Albert suggests that we do not directly perceive that space has three dimensions. Rather, we see that there are three independent directions along which ordinary objects can approach one another and interact (and so there will be three different dimensions implicit in the dynamical laws governing these motions) and infer from this that ordinary space is three-dimensional.

Why emphasize the dynamical laws? Because physics is first and foremost about how and why physical objects move around and interact with one another, and the dynamical laws are generalizations describing this behavior. We thus infer the fundamental nature of reality from the dynamical laws, which are themselves inferred from the observable behaviors of physical objects. We posit, at the fundamental level, whatever is required for the laws governing objects' motions.

We are familiar with using this sort of reasoning—the reasoning that takes us from dynamical laws to the fundamental nature of the world—for space-time. How do we infer the space-time structure of a world, according to its fundamental physics? We look at the dynamical laws and infer the structure that is needed to support the laws—"support" in the sense that the laws presuppose that structure; they cannot be formulated without assuming this structure. That is, we look at the mathematical space-time structure needed to formulate the theory and infer the corresponding physical geometry to the space-time of the world. In particular, we look at the dynamical laws formulated in a coordinate-independent, geometric way (coordinate-dependent formulations can sneak in extra structure that isn't really required) and consider the space-time structure needed to formulate the laws in this way. We then infer that this structure exists in a world governed by those laws. If the laws cannot be formulated without referring to some structure, then plausibly the structure must exist in a world governed by those laws.

In a classical mechanical world, for example, we infer that space-time is Galilean, not Aristotelian. Aristotelian space-time has all the structure of Galilean space-time, plus an additional preferred rest frame structure. Yet that further structure isn't needed or referred to by the dynamical laws: the laws are the same regardless of choice of inertial frame. This means that we can formulate the laws without assuming a preferred rest frame. So we infer that the space-time of the theory doesn't have this extra structure. If the laws were not invariant under changes in frame, on the other hand, then we would infer this structure, for the laws couldn't be formulated without it. (Compare: if the dynamical laws weren't invariant under space translations, we would infer that space has a preferred location. The laws could not be formulated without presupposing this.)

The rule to infer the space-time structure needed for the dynamical laws comes in two parts. First, we don't infer more space-time structure than what's needed for the dynamics—just as we don't infer Aristotelian space-time in a classical world. We infer the least space-time structure to the world that's needed to formulate the fundamental dynamics. Any additional structure is excess, superfluous structure, not in the world—as a choice of inertial frame in classical mechanics is an arbitrary choice in description, not a distinction in the world. Second, we infer at least as much structure as needed for the dynamics. We do not infer less than Galilean space-time structure in a classical world. The dynamical laws presuppose the distinction between straight and curved space-time trajectories, for instance, a distinction that is supported by Galilean space-time.[6]

Thus, we infer the minimal structure required for the dynamics—we adhere to a minimize structure principle[7]—and we also infer at least as much structure as required—we adhere to a "don't eliminate too much structure" principle. In other words, we infer *just that* space-time structure required or presupposed by the fundamental dynamical laws. The idea behind this principle is intuitive: a match in structure between the dynamical laws and the world is evidence that we have inferred the *correct* space-time structure to a world governed by those laws.[8]

Notice that the same intuitive idea supports a principle to posit just that fundamental structure to the world—no more, no less—as needed to support the dynamical laws, regardless of whether this is the structure of an ordinary low-dimensional space(time) or some higher-dimensional space. Regardless, the match in structure between the dynamics and the world is evidence of our having inferred the correct structure to the world. Regardless, this principle stems from the very general idea that the dynamical laws are a guide to what is fundamental to a world.

Finally, we also infer the fundamental ontology presupposed by the dynamical laws. This may be less immediately familiar, but it is something we typically adhere to. Think of Newtonian physics. In addition to indicating the space-time structure, the dynamical laws tell us that, fundamentally, there are particles, which travel along straight paths unless acted on by a net external force. The dynamics presupposes that there are such things; the laws wouldn't be true if there weren't. So we infer, in a Newtonian world, that particles exist at the fundamental level.[9] Just as a match in structure between dynamics and world indicates that we have inferred the correct structure to the world, so a match in ontology between dynamics and world indicates that we have inferred the correct fundamental ontology to the world.

In sum: we adhere to a general principle to infer just that fundamental structure and ontology that is required by the dynamical laws.

---

[6] In North 2012, I argue that classical space-time has a somewhat different fundamental structure than standardly supposed, but one that includes a fundamental affine (inertial, straight-line) structure.

[7] I argue for this principle in North 2009.

[8] Compare Earman 1986, p. 26; 1989, p. 46.

[9] This isn't to say that Newtonian mechanics could not hold in a world of which matter is fundamentally gunky, or a world fundamentally containing only macroscopic objects. In the latter case, the objects can be treated as composed of point-sized bits of matter to which the laws apply, even if there fundamentally are no particles; alternatively, the laws can be interpreted as governing the objects' centers of mass. In the former, it's unclear whether to consider the world genuinely Newtonian, though a version of Newton's laws can still hold. In typical Newtonian worlds, however, we make the inference to particles.

Three final notes on this principle. First, it is a guiding methodological principle; it will not yield conclusive inferences. We cannot be certain that the structure and ontology indicated by the dynamics is the correct structure and ontology of the world. We cannot be certain that there is no preferred frame in a classical world, for instance. Still, the principle is a reasonable guide, which we think has been successful. (Consider the inference to Minkowski space-time in special relativity. The special relativistic laws can be formulated without assuming absolute simultaneity, so it is reasonable to infer that there is no such structure in the world—reasonable, even though there could still be a preferred frame.)

Second, the form of the laws from which we read off the structure and ontology should be geometric. This is typically the simplest, most objective (coordinate-independent) statement of the laws, and thus the best guide to the nature of the world, apart from our descriptions of it.

Third, this principle applies to the fundamental level or supervenience base. It says to infer just that fundamental structure and ontology needed for the dynamics.[10]

In all, we infer the fundamental stuff the dynamical laws need in order to be geometrically formulated. In the next two sections, I suggest that we can use this general principle to figure out the fundamental space of a quantum mechanical world. And we can do so in a way that is less controversial than it might initially seem, given the principle's more familiar applications.

2 The Fundamental Space of a Quantum World

It's now a relatively short step to the conclusion that the wave function's space is fundamental to a quantum mechanical world.

In quantum mechanics, the wave function is the mathematical object that represents the state of a system at a time.[11] Think of it as like a function and like a wave, as it is called. As with a function, the wave function takes in points of the space on which it's defined and gives out values, here complex numbers. As with a wave (or field), the wave function assigns a (complex) number, or *amplitude* (a "height"),[12] to each point in the space in which it lives. On the standard view, the wave function represents everything about the fundamental state of a system (where this system could be the entire world) at a time. (In Bohm's

---

[10] We can think of this as an updated version of Quinean ontological commitment. Not: what there is, is what the values of the variables range over, so that we first render our theory in (first-order) logic and then see what the values of the variables are. Rather, what there fundamentally is, is given by the (best invariant formulation of the) dynamical laws, so that we first render our fundamental theory in geometric terms and then infer the structure and ontology presupposed by the laws.

[11] Alternatively, we can use a (normalized) vector (or a ray). I discuss later why we can set aside this other mathematical formulation here.

[12] And phase, which I ignore here for convenience.

theory, the fundamental state is given by the wave function plus the positions of a system's particles.) A system's history is given by the evolution of its wave function over time, in accord with the dynamical laws. These laws include (at least) the deterministic Schrödinger equation. (Depending on the theory, there may be other fundamental dynamical laws; see note 4. Ignore these complications here.)

The space on which the wave function is defined is high-dimensional: $3n$ dimensions for a world containing what we ordinarily think of as n particles in three-dimensional space.[13] (Whether there really are particles depends on the theory; more on this later.) This space, which I have been calling the wave function's space, is similar to what is called "configuration space," but these should not be confused. A configuration space represents ordinary particle configurations. Think of classical mechanics, where the configuration space of an n-particle system has $3n$ dimensions, one for the location of each particle along each of three ordinary spatial dimensions; each point in this high-dimensional space represents a possible configuration of particles in three-space. A quantum mechanical configuration space is similar in that each point represents a configuration of particles in ordinary three-dimensional space. In short, configuration space fundamentally represents particle configurations in three-space.

The view I defend is that the wave function's space is fundamental. Unlike configuration space, this high-dimensional space doesn't fundamentally represent particle configurations in three-dimensional space; the structure of this space isn't given by particle positions in three-space. The wave function's space, not three-space, is the fundamental space here. The wave function's space is isomorphic to configuration space, but it should not be confused with a genuine *configuration* space, in the sense of a space that fundamentally represents particle configurations in three-dimensional space. Fundamentally, there is no three-space on this view; a fortiori, fundamentally, there are no particles in three-space.

Now, in a theory like classical mechanics, configuration space is seen as just a mathematical tool.[14] The dynamics can be formulated either on the high-dimensional configuration space or a three-dimensional space. Ordinary experience then suggests that the three-dimensional space is the one that accurately represents the world's fundamental physical space.

In quantum mechanics, however, we must formulate the dynamics on a high-dimensional space. This is because quantum mechanical systems can be in entangled states, for which the wave function is nonseparable. Such a wave function cannot be broken down into individual three-dimensional wave functions,

---

[13] I leave out spin for convenience. Spin can be represented by extra internal degrees of freedom at each point of the wave function's space.

[14] But see North 2009, 2012 against its being merely a mathematical tool.

corresponding to what we think of as particles in three-dimensional space. That would leave out information about correlations among different parts of the system, correlations that have experimentally observed effects. Only the entire wave function, defined over the entire high-dimensional space, contains all the information that factors into the future evolution of quantum mechanical systems.[15]

Following the principle to infer, at the fundamental level of the world, just that structure and ontology that is presupposed by the dynamics, we are led to conclude that the fundamental space of a world governed by this dynamics is the high-dimensional one. The fundamental ontology, which includes the wave function, then lives in it.[16] (Note that, on a wave function space version of Bohm's theory, the fundamental ontology also includes a "world particle," whose dynamical evolution in the wave function's space gives the evolution of the [nonfundamental; see section 4] "particles" in three-space.)

Of course, it is important to keep in mind the distinction between the wave function as a mathematical object and as a real physical field—likewise, between the abstract space on which the mathematical object is defined, and the physical space on which the physical field lives. So why not take the wave function and its space as mathematical tools that do not represent physical things in the world? Because of our guiding principle. This principle says to infer, from the mathematical structure needed to formulate the dynamical laws, the corresponding physical structure and ontology to the world. Compare the space-time case, in which we take a theory's mathematical structure seriously in that it corresponds to a particular physical geometry in the space(time) of the world.

(Another standard formulation of quantum mechanics uses a different mathematical space—an abstract vector space called Hilbert space. Unlike the wave function's space, though, this space is not a candidate fundamental physical space of a quantum world: Hilbert space is just a mathematical tool that yields a convenient formulation of the theory. Our general principle arguably yields this result. Recall that part of this principle warns against inferring too little fundamental structure to the world. One way to infer too little structure is by positing too minimal a basis on which to recover the ordinary world of our experience. The Hilbert space formulation seems to contain too little structure from which to construct a picture of the world as we experience it. Hilbert space does not support an objective, structural distinction between positions and other physical

---

[15] Thus (here speaking in terms of particles, though remember that on some views fundamentally there are no such things), two particles' locations might be perfectly correlated (always in the same region) or anticorrelated (in different regions). When projected onto the three-dimensional space for each particle, the wave function for such an entangled system looks the same whether the particles are correlated or anticorrelated. See Lewis 2004, section 2; Ney 2010a, section 3.3, Ney 2010b; Lewis's chapter in this volume for this argument.

[16] Or, since on this view the wave function is a physical field, it may be better to say that it lives "on" this space; Maudlin 2010, p. 126.

properties, like spin, in the way that the wave function's space does.[17] In addition, the fundamental objects of Hilbert space are vectors; the wave function space formulation, on the other hand, allows for a more familiar particle and field ontology [albeit a fundamentally very high-dimensional such ontology]. Hilbert space is best interpreted as an abstract statespace; whereas the wave function's space can be interpreted as a physical space, inhabited by the fundamental physical objects of the theory. In all, it is too hard to recover a perspicuous picture of the world from the Hilbert space formalism. Of course, what counts as perspicuous is a matter of debate. Wallace and Timpson [2010] [and in a different way Maudlin 2007, 2010] argue that wave function realism does not yield a perspicuous theory. I disagree, for reasons given in section 4.)

3 Against Fundamental Three-Space Views

Our general principle tells against views which maintain that a three-dimensional space is fundamental to a quantum world. (Note that on any such view, the high-dimensional space of the wave function is a genuine configuration space.)

One view says there are two fundamental physical spaces, ordinary three-dimensional space and a high-dimensional configuration space. The wave function, in configuration space, governs the motions of particles (or other objects, such as mass densities or flashes [Allori et al. 2008]; there will be some such fundamental objects, on this view) in three-space.

This theory has more structure than any view positing a single fundamental space. For there are two distinct fundamental spaces, each with its own structure. What's more, each space must possess additional structure beyond what is normally attributed to it. Further structure is needed to ground the connections between the two fundamental spaces, saying which parts and dimensions of the high-dimensional space correspond to which parts and dimensions of ordinary space, and which axes of configuration space correspond to which particle. Notice that this is additional *fundamental* structure. As such, it goes against our principle. This is extra fundamental structure beyond what is needed for the dynamics; it is excess structure we should do without.[18]

---

[17] Compare Wallace and Timpson 2010, p. 703: "the physical universe is ... very highly structured, whereas Hilbert-space vectors seem pretty much alike."

[18] David Albert (in a seminar of Tim Maudlin's at Rutgers University in 2007) has raised a similar concern, arguing against what he calls the additional "metaphysical structure." Dorr 2009 is a version of the view. (See also the "mixed ontology" view of Monton 2002.) Dorr defends realism about a fundamental configuration space and a fundamental three-space, with fundamental "putting" relations connecting the two. It is hard to compare overall structure here: do further fundamental relations add fundamental structure? It seems to me that this will require more structure than any single-fundamental-space view, though I admit that this isn't clear-cut.

David Albert has noted[19] that the laws will also be odd on such a view. Ordinarily, we predict how things will behave by means of direct geometric relations among them. That is, all physical objects inhabit a single space, and the laws governing their motions and interactions are formulated in terms of the geometry of that space. Thus, consider the electromagnetic field and charged particles in classical electromagnetic theory. The direction and magnitude of the field everywhere determines, in accord with the dynamical laws, how the charges in various configurations will move around. We can look at the one space, figure out what direction the field is pointing in at a given location, and, using the laws, infer that a test particle will head in that direction when it passes through the location. This is unlike the two-fundamental-spaces view of quantum mechanics, in which the two kinds of thing in the world occupy distinct spaces, and so lack any direct geometric relations between them; the laws relate the wave function, in one space, to the particles, in another. (This is where the extra fundamental structure comes in, to say that *this* direction in configuration space corresponds to *that* direction in physical space, indicating how the wave function's behavior in its space causes the particles to move around in their space.) It's odd for fundamental laws to be formulated in terms of structure connecting distinct spaces, instead of the intrinsic geometry of a single space that everything inhabits. This is unlike other physical theories with which we are familiar, and it obfuscates our understanding of how the different objects interact.

The minimize structure principle says to infer that one space alone is fundamental. Since the dynamics requires the high-dimensional space of the wave function, we should infer that this space represents the fundamental physical space of a world governed by that dynamics. The defender of the two-fundamental-spaces view will reply that ridding the world of three-space is ridding the world of too much structure, against the "don't eliminate too much structure" principle. Yet doing away with a fundamental three-space isn't yet to say that no such space exists: there could still be a nonfundamental three-space. I discuss this in section 4.

(Why does the Hilbert space formulation abstract away too much, whereas a fundamental wave function space view does not? I have no conclusive reason for this. The thought is something like the following. It isn't abstracting away too much if you do not privilege the existence of ordinary objects at the fundamental level [see section 4]; we learned that with the advent of atomic physics. But you should have room in your fundamental theory for drawing certain ordinary distinctions in some way or other; doing without such distinctions at the fundamental level would be abstracting away too much. Exactly how these distinctions are metaphysically accommodated does not matter. But it matters that you make some room for them in your fundamental theory. Otherwise, it will be

---

[19] At a conference at Rutgers University in 2007.

too difficult to construct an ordinary picture of the world on the basis of the theory. Thus, the Hilbert space formulation abstracts away too much because it doesn't make room at the fundamental level for ordinary notions like position—it denies an objective distinction between, for example, spin facts and position facts—and it doesn't allow for a fundamental particle-field distinction. The wave function space formulation, on the other hand, does not abstract away too much by denying that ordinary objects exist at the fundamental level. More generally, there are certain kinds of facts without which a fundamental physical theory abstracts away too much, and certain other kinds of facts without which it doesn't. The best theory posits the structure and ontology required by the dynamics while allowing for enough ordinary facts and distinctions. This is admittedly vague, but I hope somewhat intuitive.[20])

Other views say that three-space alone is fundamental. Bradley Monton (2006, this volume) argues that quantum mechanics is fundamentally about particles in three-space. The wave function doesn't "live" on a physical space. It is a mathematical tool, defined on an abstract configuration space, which represents the quantum mechanical properties of ordinary particles.

This view faces a dilemma. Either Monton says that the quantum mechanical laws are about the wave function in configuration space, in which case he violates the principle that the dynamical laws are about what's fundamental, since for him the wave function and configuration space aren't fundamental. Or he says that the laws are about ordinary particles' properties, which are fundamental. But in that case the laws will likely be very complicated. (I say "likely" because he doesn't say exactly how the different quantum mechanical properties of particles are related to one another.) To be stated solely in terms of things that Monton takes to be fundamental, the laws must be formulated as constraints on ordinary particles' properties, like their locations in three-space. Yet it is hard to see how the quantum dynamics can be simply formulated in this way. The version of the laws we are familiar with employs the geometry of the high-dimensional space of the wave function. Monton's view also leaves us with a nagging question: what is it about particles' properties such that the abstract configuration space description is the *right* way to represent them? In all, this view lacks structure that's needed for the simple, geometric formulation of the dynamics.

Peter Lewis (2004) argues that quantum mechanical configuration space is fundamental, but that it has *three* dimensions in a relevant sense. This space is $3n$-dimensional in requiring "that many independent coordinates to parameterize the properties of the system" (2004, p. 726). But in another sense, it is three-di-

---

[20] The same reasoning deems the Schrödinger picture, and not the Heisenberg one, a candidate for describing the fundamental nature of a quantum world, despite the mathematical equivalence. The Heisenberg picture lacks a metaphysically perspicuous picture of the world: there is just one physical state, unchanging in time; only the operators change. Maudlin 2010, pp. 128–29 discusses the ontological obscurity of this picture. Muller 1997a, 1997b discusses the equivalence between the Heisenberg and Schrödinger formalisms, and notes ways in which the resultant theories are inequivalent.

mensional: the parameters needed to describe systems' fundamental states have a preferred grouping into threes. So there is a way of understanding everything that happens as unfolding in a space with three independent spatial directions.[21]

This does not alleviate the main problem for configuration space (or wave function space) realism, though. Even if configuration space is three-dimensional in some abstract, representational way—even if there is an abstract way of capturing what goes on in three dimensions—intuitively, the space of the theory is still *really*—fundamentally—3n-dimensional, for this number of dimensions is needed to formulate the dynamical laws.[22] The problem for wave function space realism stems from the fundamental geometry of this space, not the mathematical geometry of spaces we can use to represent it. The problem is that the fundamental geometry needed to formulate the theory, and the corresponding physical geometry we infer to the world, is not three-dimensional, contrary to what our experience suggests. That problem remains. (Ultimately, there may not be a deep disagreement here. I also think there is a sense in which a quantum world is three-dimensional: there exists a nonfundamental three-space [see section 4]. But I disagree that the wave function's space itself is three-dimensional in any sense; nor do I think that a quantum world is fundamentally three-dimensional. Whether there is a real disagreement depends on whether Lewis would agree with me that there is such a thing as the world's fundamental geometry, which can differ from the geometry of a nonfundamental space.)

Another view maintaining the fundamentality of three-space is Tim Maudlin's (2007).[23] According to Maudlin, configuration space is a mathematical tool for defining the wave function, which governs the behavior of ordinary particles.[24] The wave function in configuration space—alongside a world particle, in Bohm's theory—yields, in Maudlin's term, an *informationally complete* description, from which "every physical fact about the situation can be recovered" (2007, p. 3151). But this isn't an *ontologically complete*, "exact representation of all the physical entities and states that exist" (2007, p. 3154). In other words, the wave function in configuration space is not an *ontologically accurate* (my phrase, not Maudlin's[25])

---

[21] Lewis (in this volume) goes on to argue that a quantum world is really three-dimensional—that the sense in which configuration space is 3n-dimensional is misleading, and in any case non-spatial.

[22] In other words, the high-dimensional space is needed for a theory that's "dynamically complete," in the sense to be discussed shortly.

[23] Allori et al. (2008) (see also Allori in this volume) is a relevantly similar view, though one that disagrees on the status of the wave function. Allori et al. (see also Goldstein and Zanghì in this volume) suggest it is like a law (though its precise status depends on the particular theory of quantum mechanics); Maudlin (2007) refrains from putting the wave function in a particular category. In a talk I heard some years ago, Maudlin suggested that the wave function is unlike anything else, in its own metaphysical category.

[24] Or flashes or mass densities: Allori et al. 2008.

[25] I use the phrase because saying that the wave function isn't ontologically complete suggests that the wave function is in the ontology, it just isn't everything. In Maudlin's view, the wave function isn't in the physical ontology—it isn't a physical field.

depiction of a quantum world, even though it allows us to predict everything that happens and so is informationally complete. The ontologically accurate description is instead given by particles in three-space (in Bohm's theory; alternatively, mass densities or flashes in three-space in GRW[26]), even though this isn't informationally complete (since that requires the wave function).

In general, Maudlin warns, we cannot assume that informationally complete descriptions are ontologically accurate. If we did, then we would eliminate charged particles from the ontology of classical electromagnetism, for instance, since there is an informationally complete description without them (because the divergence of the electric field suffices to give the charges' locations). Likewise, for any deterministic world, we would posit only whatever is in the world at one time, since the state at a time plus the laws is informationally complete. In the case of quantum mechanics, Maudlin argues, although the wave function in configuration space is needed for informational completeness, there are independent reasons—namely, the evidence from ordinary experience—for positing three-dimensional objects, not the wave function, in the ontology.[27]

This brings us to a basic disagreement between wave function space and ordinary space views: how much to emphasize the dynamics in figuring out the fundamental nature of the world. Three-space views prioritize our evidence from ordinary experience, claiming that the world appears three-dimensional because it is fundamentally three-dimensional. Wave function space views prioritize our inferences from the dynamics, claiming that the world is fundamentally high-dimensional because the dynamical laws indicate that it is. Notice that, although this latter view is counterintuitive, there is precedent for the inference it relies on, as in the case of space-time structure discussed earlier. Indeed, we can rely on a similar inference for the ontology of classical electromagnetism, too. Although the field values will give the locations of charges, as David Albert has noted,[28] this will not give their masses, which are also needed to predict particle and field value locations at other times. In other words, the field values aren't, in Albert's phrase, *dynamically complete*. That's why we do not eliminate charges from the ontology: they are required by the dynamical laws. (If the field description were dynamically complete, on the other hand, we might well conclude that charges aren't in the ontology.[29]) Wave function space views hew to this tradition of positing, in the fundamental level of the world, whatever is required by the dynamics.

Against Maudlin, then, I think that informational completeness of the right sort—dynamical completeness—does track ontological accuracy. A dynamically

---

[26] Though Maudlin 2010 gives considerations against the mass density picture.

[27] There can be informationally complete descriptions that are ontologically complete, like particles in classical mechanics. Maudlin suggests that whether informational completeness and ontological completeness apply to the same description is to be decided on a case-by-case basis.

[28] In a seminar at Columbia University in 2008.

[29] There was a history of trying to do this, but it didn't work: Arntzenius (1993).

complete description contains the structure and ontology presupposed by the fundamental dynamical laws.[30] For Maudlin, dynamical completeness and ontological accuracy can come apart: in quantum mechanics, there is dynamical information that doesn't correspond to any structure in the physical world. According to our general principle, however, the fundamental level of the world—the fundamental ontology, the fundamental space, and its structure—should contain whatever is required to formulate the dynamical laws; there should be a match between the structure needed for the dynamics and the fundamental structure of the physical world. If so, then we should infer that the fundamental physical space of a quantum world is the high-dimensional space of the wave function, with the fundamental ontology residing in it.

4 The Structure of Appearances

Assuming that our world is fundamentally quantum mechanical, there remains the question of whether wave function space realism can explain the fact that we appear to live in three-dimensional space.

Maudlin (2007) argues that it can't, because it lacks what J. S. Bell calls "local beables" (1987, pp. 52–53), parts of the ontology localized to regions of ordinary space(time). Indeed, Maudlin says, it is hard to see how we could ever come to understand, let alone empirically confirm, such a theory, when all of our evidence takes the form of local beables.[31]

Now, it is true that there are no *fundamental* local beables, on this view. But this doesn't mean that there are no nonfundamental such things. Indeed, I think that something like this holds for three-space and its objects as a whole. Unlike Albert (1996), who argues that in a quantum world, ordinary space is an "illusion" and our talk about it is false,[32] I think that three-space *exists* in such a world, and our talk about it is true. It's just that this space is *nonfundamental*. Similarly, ordinary particles exist but are nonfundamental. They are more like tables and chairs: made up out of fundamental stuff, not themselves in the fundamental inventory.[33]

In place of Albert's antirealism about three-space, I suggest an antifundamentalism. Ordinary space exists at a "higher level." Even so, there are objective facts about it and we can say true things about it. It's just that none of this is fundamental—just as tables and chairs exist and have objective truths about them but are not fundamental. We might say that statements about these things aren't *strictly speaking* true, but this just means that they are not *fundamentally*

---

[30] On this understanding, the state at one time in a deterministic world is not dynamically complete. For the laws relate states at different times, thereby presupposing that there are such states.

[31] See Maudlin 2010 for further argument along these lines.

[32] Ney 2010b in a different way argues that three-dimensional space doesn't exist in a quantum world.

[33] Compare Albert and Loewer 1995 and Wallace 2003, 2010.

true. There is an objective fact as to where something is located in three-space, even though there is no such fundamental fact.

There are tricky issues here about exactly how to understand the claim that ordinary space exists but is nonfundamental. On some recent views in metaphysics, we cannot make sense of such claims; we must say that the nonfundamental things simply do not exist.

I think that there is a way of making sense of the idea that ordinary space is nonfundamental yet real, in the same way that ordinary objects, the special sciences, and so on, are nonfundamental yet real. A *grounding relation*[34] captures the way that the wave function's space is fundamental and ultimately responsible for ordinary space, while at the same time allowing for the reality of ordinary space. This is an explanatory relation that captures the way in which one thing depends on or holds in virtue of another, without implying that the dependent thing does not exist. Thus, three-dimensional space and its objects are *grounded in* the wave function's space and its objects. For example, there being a table in three-space consists in nothing but the wave function's having a certain shape in its high-dimensional space. It's *true* that there is a table in three-space; it's just that this holds *in virtue of* some other, more fundamental facts. The truth about three-space (the grounded) is not a further fact beyond the truth about the wave function's space (the grounds)—that is, it isn't a fundamental fact—even though it is distinct from the grounds and is itself a real fact.

More generally, the wave function's space is fundamental, and three-space is grounded in it; what's true of three-space holds in virtue of what's true of the wave function's space. This captures the way that three-space is emergent but "no less real for that" (Wallace and Timpson 2010, p. 706). It also captures the idea that three-dimensional happenings *are* nothing over and above various wave function space happenings; that is, that three-space is not fundamental. In the way that thermodynamic or biological happenings, say, are nothing over and above various particle happenings—the former processes are grounded in more fundamental particle processes—so, too, for ordinary three-space happenings vis-à-vis what goes on in the wave function's space. (Thus, the grounding relation more generally captures the way in which there are ordinary macro-level sciences, with generalizations that are objectively true. They just aren't fundamental, but hold in virtue of what goes on at the fundamental level.)

I submit that this is the overall simplest, empirically adequate account of a quantum world. It explains our experience and captures the truth of our ordinary claims about three-dimensional space, while at the same time positing just that structure that's needed for the dynamics.[35] This view has the benefit of

---

[34] Say, like that of Fine 2001; what I say here is neutral on the metaphysics of grounding.

[35] Fine 2001, p. 22 notes that we can evaluate a system of grounds "in much the same way as any other explanatory scheme, on the basis of such considerations as simplicity, breadth, coherence, or non-circularity" and most importantly, "explanatory strength."

a fundamental three-space view—what explains the fact that the world appears three-dimensional is that there exists a three-dimensional space—while also matching the structure and ontology for the dynamics. (What it doesn't do is explain the fact that three-space appears fundamental by saying that three-space is fundamental.) In particular, there is no fundamental structure beyond what is needed for the dynamics. For there is no fundamental structure connecting ordinary space and the wave function's space; there are simply "grounding rules" from the fundamental to the nonfundamental, and these do not add fundamental structure, in the same way that correspondence rules for the special sciences don't add fundamental structure to the world. This picture also avoids the worry raised by Monton (2006) that the view is radically revisionary. It is indeed fundamentally revisionary, but it is not revisionary about the nonfundamental.

You might wonder why the wave function's space grounds an emergent three-dimensional space, not some other. Monton (2006) argues that it doesn't manage to do this, because there is no intrinsic structure in the wave function's space marking out a preferred grouping of axes into threes. There is nothing special about the fact that the number of dimensions is equal to "3 times n": this space isn't fundamentally about particles in three-space.

But we have more to work with than just the kinematical structure of the space. There is also the dynamical structure, and this opens avenues of response. Lewis (2004) (see also Lewis's chapter in this volume) argues that the wave function's space does have intrinsic structure picking out a preferred grouping of dimensions into threes. Albert (1996) alternatively suggests that the form of the Hamiltonian results in the illusion of three-space, without extra intrinsic structure. Albert argues that the Hamiltonian has a uniquely natural form in three dimensions; in my view, this naturalness is evidence that the grounded space is, in fact, three-dimensional. Wallace and Timpson (2010) agree with Albert's point, adding that the experience of three dimensions should emerge due to decoherence.[36]

I suspect that one of these views is correct. One of these can explain how the fundamental facts about the wave function (and perhaps a world particle) in the high-dimensional space ground the three-dimensional facts. Even if not, though, we could take this as some additional fundamental structure. This structure isn't needed for the dynamics. But there is more guiding theory choice than just the "posit what's needed for the dynamics" rule. There is also empirical adequacy, which may require a primitive "preferred grouping of axes into threes" structure. That is, the wave function's space itself may have an additional level of structure marking where its dimensions group themselves into threes; this would then ground the three-space facts. Even so, this view

---

[36] Though they argue that this is insufficient to fully recover the world.

is preferable overall. For it has just about the fundamental structure needed for the dynamics, while also explaining how the fundamental facts ground the three-dimensional facts. Notice that on any of these approaches, this way of grouping the dimensions of the fundamental space is the *right* way of doing so, because it captures the truths about the nonfundamental—just as there is a correct way of carving up the fundamental statespace into macroscopic parameters, namely, the way that yields the truths about the higher-level sciences. Of course, none of this is to say exactly how the grounding of three-space in the high-dimensional space occurs; that question remains. But it answers the objection from local beables.

Finally, you might worry that the structure of the wave function's space remains unexplained on this view. I cannot say that it has the structure and dimensionality it does because it represents particle configurations in three-dimensional space. It isn't fundamentally about particles in three-space.

According to my view, however, the structure of the wave function's space is fundamental, not in need of explanation on the basis of anything more fundamental. We infer this structure from other things, like the dynamical laws; but this space has the structure it does because the world fundamentally is the way that it is. It may seem remarkable that it has just the right structure to yield the appearance—and the existence—of a three-dimensional space. But of course it does, if this really is the fundamental theory and those are the appearances that the theory saves.

Think of it this way. The relation between the wave function's space and its ontology, on the one hand, and three-dimensional space and its ontology, on the other, is analogous to the relation between particles, on the one hand, and tables and chairs, on the other. Compare: isn't it remarkable, if particles are fundamental, that they should conspire to make it seem as though there really are tables and chairs? But of course particles conspire to form themselves into tables and chairs, if particles really are in the fundamental level of reality and the nonfundamental stuff includes tables and chairs. Since the apparent existence of tables and chairs is the starting point for our theorizing, of course the fundamental theory we are led to is one that predicts the appearances (and existence) of tables and chairs. To put it another way, our *evidence* for the theory, in the first place, is what we observe. But what we observe, everyone agrees, is a parochial reflection of our own situation: we are familiar with tables and chairs. It is then no great coincidence that we end up with a fundamental theory that has the power to predict the appearances for us.

5 Conclusion

Why conclude that wave function space realism (or wave function space fundamentalism) is a physically accurate picture of a quantum world? Why not think the wave function's space is just part of the mathematics used to formulate the theory? Because we generally posit, in the physical world, the fundamental structure and ontology presupposed by the dynamical laws. This match between dynamics and world is evidence that this *is* the fundamental nature of a world governed by that dynamics.

References

Albert, David Z. (1996). "Elementary Quantum Metaphysics." In J. T. Cushing, A. Fine, and S. Goldstein, eds., *Bohmian Mechanics and Quantum Theory: An Appraisal*, pp. 277–84. Dordrecht: Kluwer.

Albert, David, and Barry Loewer (1995). "Tails of Schrödinger's Cat." In R. Clifton, ed., *Perspectives on Quantum Reality*, pp. 181–92. Dordrecht: Kluwer.

Allori, Valia, Sheldon Goldstein, Roderich Tumulka, and Nino Zanghì (2008). "On the Common Structure of Bohmian Mechanics and the Ghirardi-Rimini-Weber Theory." *British Journal for the Philosophy of Science* 59(3):353–89.

Arntzenius, Frank (1993). "The Classical Failure to Account for Electromagnetic Arrows of Time." In T. Horowitz and A. Janis, eds., *Scientific Failure*, pp. 29–48. Lanham, Md.: Rowman & Littlefield Savage.

Bell, John S. (1987). *Speakable and Unspeakable in Quantum Mechanics*. Cambridge: Cambridge University Press.

Dorr, Cian (2009). "Finding Ordinary Objects in the World of Quantum Mechanics." Unpublished manuscript.

Earman, John (1986). *A Primer on Determinism*, vol. 32. University of Western Ontario Series in the Philosophy of Science. Dordrecht: Reidel.

Earman, John (1989). *World Enough and Space-Time*. Cambridge, Mass.: MIT Press.

Fine, Kit (2001). "The Question of Realism." *Philosopher's Imprint* 1.

Lewis, Peter (2004). "Life in Configuration Space." *British Journal for the Philosophy of Science* 55:713–29.

Maudlin, Tim (2007). "Completeness, Supervenience, and Ontology." *Journal of Physics A: Mathematical and General* 40:3151–71.

Maudlin, Tim (2010). "Can the World Be Only Wavefunction?" In Simon Saunders, Jonathan Barrett, Adrian Kent, and David Wallace, eds., *Many Worlds?: Everett, Quantum Theory, and Reality*, pp. 121–43. Oxford: Oxford University Press.

Monton, Bradley (2002). "Wave Function Ontology." *Synthese* 130:265–77.

Monton, Bradley (2006). "Quantum Mechanics and 3N-Dimensional Space." *Philosophy of Science* 73(5):778–89.

Muller, F. A. (1997a). "The Equivalence Myth of Quantum Mechanics—Part I." *Studies in History and Philosophy of Modern Physics* 28(1):35–61.

Muller, F. A. (1997b). "The Equivalence Myth of Quantum Mechanics—Part II." *Studies in History and Philosophy of Modern Physics* 28(2):219–47.

Ney, Alyssa (2010a). "Are There Fundamental Intrinsic Properties?" In Allan Hazlett, ed., *New Waves in Metaphysics*, pp. 219–39. New York: Palgrave Macmillan.

Ney, Alyssa (2010b). "The Status of our Ordinary Three Dimensions in a Quantum Universe." *Noûs*. doi: 10.1111/j.1468-0068.2010.00797.x.

North, Jill (2009). "The 'Structure' of Physics: A Case Study." *Journal of Philosophy* 106(2):57–88.

North, Jill (2012). "Structure in Classical Mechanics." Unpublished manuscript. Available at https://courses.cit.cornell.edu/north/.

Wallace, David (2003). "Everett and Structure." *Studies in History and Philosophy of Modern Physics* 34:87–105.

Wallace, David (2010). "Decoherence and Ontology." In Simon Saunders, Jonathan Barrett, Adrian Kent, and David Wallace, eds., *Many Worlds?: Everett, Quantum Theory, and Reality*, pp. 53–72. Oxford: Oxford University Press.

Wallace, David, and Christopher G. Timpson (2010). "Quantum Mechanics on Spacetime I: Spacetime State Realism." *British Journal for the Philosophy of Science* 61:697–727.

A Prolegomenon to the Ontology of the Everett Interpretation*

DAVID WALLACE

This chapter began as an attempt to explain, without assuming any technical physics or advanced mathematics, the metaphysical (and in particular, the ontological) implications of the Everett interpretation of quantum theory, often known as the many-worlds interpretation.

The difficulty is that physics is not, by and large, *gratuitously* technical. It is possible to give a nontechnical impression of what Everettian quantum theory is like, but metaphysicians don't just need impressions: they need clear statements of just what a given physical theory is saying.

In my view—and this is an underlying theme of the chapter—this has led to trouble. Physical theories don't, straightforwardly, *say* anything: the theory, shorn of any interpretation, is a piece of mathematics. In the case of classical point particle mechanics—that is, in the case of Newton's physics—that piece of mathematics is relatively easy to understand intuitively, at least at a surface level. But in the case of quantum theory, the mathematics is very abstract, and its connection to observable facts is both indirect and controversial. There is a great temptation, in trying to communicate quantum physics to philosophers, to tacitly smuggle in controversial interpretive posits as if they were part of the formalism; there is a comparable temptation to present certain, relatively simply describable special cases as if they were general; there is, in short, a serious danger of miscommunication, and this is exacerbated by the very different styles of philosophy and modern physics, a difference not entirely recognized by philosophers whose model of physics remains something like Newton's laws.

In this chapter, I do not pretend to present the mathematics of quantum theory more carefully, in a way that avoids any such communications failure; such a task lies far beyond its scope. My goal, instead, is to give some insight into the features of quantum mechanics, mathematically and interpretationally, that must

* Thanks to Chris Timpson for helpful comments on this chapter.

be understood before any metaphysical discussion of the Everett interpretation can get off the ground. Much of what has been said so far on the subject, I believe, has failed to fully understand those features; as such, it has begun the discussion on the wrong foot. By the end of the chapter, I will have begun a positive discussion of the Everett interpretation's metaphysical consequences, but I can do no more than begin it without presuming a good deal of technical physics. (As it is, I try to avoid assuming anything more complicated than complex numbers.)

The structure of the article is as follows. In section 1, I explain the concept of *state space,* a commonplace idea in physics since at least the nineteenth century but one which will be unfamiliar to most philosophers (though it has a family resemblance to the set of possible worlds). In section 2 I apply this discussion to quantum mechanics, and in doing so provide a perhaps unfamiliar perspective on the quantum measurement problem. In section 3 I outline the Everett interpretation and its claims about higher-level ontology: about chairs, tables, and the like. In sections 4 and 5 I consider the question of fundamental ontology in the Everett interpretation, first exhibiting what I think are the main misconceptions available, and then making some provisional comments on what positive story can be told.

1 The Concept of State Space

Perhaps the most familiar theory in physics is the theory of Newtonian point particles. Physically, the theory is about the movement of objects through space, where those objects are small enough relative to the distances between them that they can be treated as point-like. (The mechanics of the planets in the solar system is the most important practical example.) Mathematically, we represent the instantaneous state of the system of particles by a collection of points in three-dimensional space.

It is perfectly possible to do the mathematics of Newtonian mechanics in terms of this collection of points, but more advanced treatments of mechanics instead represent the instantaneous state of an N-particle system not by N points in three-dimensional space but by one point in $3N$-dimensional space— with the first three coordinates encoding the position of particle 1, the second three encoding the position of particle 2, and so forth. This $3N$-dimensional space is called the *configuration space* of the system—so called because every point in it represents a possible instantaneous configuration of the system. And so it is possible to reexpress the equations of Newtonian physics—the equations for how those N particles move through space—as equations governing the movement of a single point through configuration space.

This is just one example of a general move made throughout physics, and more generally in the study of dynamical systems: the introduction of a *state space,* where each point in the space represents the state of the entire system being studied. A curve in such a space represents an entire history of the system (one point for each moment of time).[1] A great many such state spaces are used in physics and elsewhere in science (indeed, multiple sorts of phase space are used even in classical mechanics). Not all of them regard the points in the state space as representing possible physical states of a system; sometimes, a point in the state space needs to be understood as representing the various *probabilities* (be they chance or credence) that the actual physical system in question has of being in a given state.

It is important to realize that mathematically speaking, state spaces are highly structured: they are not mere sets of points. *Some* of that structure is specified by the very general nature of the dynamical theory under study: the configuration spaces of *any* classical-mechanical theory, for instance, must be so-called differential manifolds. But *most* of the mathematical structure is fixed by the specific details of the system under study. In the theory of N Newtonian point particles, for instance, configuration space cannot be a *homogeneous* 3N-dimensional space: it has has to have enough structure that we can read off the positions of each separate particle from a single point in configuration space. (The mathematical details of this are unimportant; it suffices to observe that in a homogeneous 3N-dimensional space, no structural features of the space distinguish one point from another.)

Indeed, while a physical system represented in the state space formalism may have a huge amount of structure, that structure is not encoded at all in the state space point (which, as a mere point, is entirely structureless). *All* the structure is encoded in the location of the point in state space, which in turn is only possible because of the highly structured nature *of* state space.

Metaphysically speaking, the state space concept may seem of little relevance—the metaphysics of a theory, presumably, is to be understood in terms of the actual properties and relations holding between the objects that make up the world according to the theory, and state space is just an abstract mathematical tool. Certainly in Newtonian mechanics (and arguably in classical mechanics more generally), this seems correct: N-particle point mechanics is a theory of N particles moving in homogeneous three-dimensional space, not a theory of one particle moving in a highly inhomogeneous, very high-dimensional space, and the fact that we can represent those N particles via the state space formalism is mathematically useful but not metaphysically germane. Unfortunately, in quantum mechanics, things are not so straightforward.

2 Quantum Mechanics and the Measurement Problem

Quantum theory, unlike classical mechanics, is usually first presented in its state space form. The particular state spaces used for quantum physics are called

---

[1] Mathematically speaking, this means that a history is really a parameterized curve, or path; equivalently, it's a map from whatever space represents the instants of time into the state space.

Hilbert spaces,[2] and they come equipped with a deterministic dynamics: through every point in Hilbert space there is exactly one dynamically permissible curve. Hilbert space also has a crucial property called *linearity*: if ψ and ϕ are points in the state space, then it makes sense to talk about the state $\alpha\psi + \beta\phi$, where α and β are complex numbers satisfying $|\alpha|^2 + |\beta|^2 = 1$. Furthermore, the dynamics respects this linearity: if ψ evolves after t seconds into $\psi(t)$, and ϕ evolves after t seconds into $\phi(t)$, then $\alpha\psi + \beta\phi$ evolves after t seconds into $\alpha\psi(t) + \beta\phi(t)$.

One might expect that to understand the relation between quantum mechanics and physical reality (and in particular, to understand the way in which quantum mechanics has empirical significance), we would need to understand what physical goings-on are represented by each point in Hilbert space. But in fact this is not how the relation between quantum mechanics and the world is normally presented. Instead, textbooks generally take an explicitly instrumentalist line: to every quantum system is associated a set of possible *measurements* that can be carried out on it, some though not all of which are designated as measurements of classical quantities like position and momentum. Given any such measurement, each state in the Hilbert space uniquely determines a probability distribution over the possible outcomes of that measurement, which is interpreted as the probability that a given measurement, if actually performed, will produce a given outcome.

That this rule provides empirically adequate predictions is not in question: quantum theory is by far the most thoroughly empirically confirmed theory in science. What *is* in question is how to go beyond this rather mysterious algorithm, and understand quantum theory as a description of physical reality: as David Lewis puts it, the challenge is to "see how [quantum mechanics] looks when it is purged of instrumentalist frivolity and dares to say something not just about pointer readings but about the constitution of the world" (Lewis 1986, p. xi). This is (one way of describing) the infamous *quantum measurement problem*: to solve the problem is to either make sense of unmodified quantum mechanics as a physical theory in good philosophical standing, or to replace it with some equally empirically adequate theory in good standing. (Were this chapter aimed at a physics audience I would digress for some time as to just why any solution—any move beyond instrumentalist "frivolity"—is needed; metaphysically inclined philosophers, I take it, need no such digression.)

To see the severity of the problem (and, in doing so, to get clearer on the form of the measurement algorithm), let's attempt to make sense of quantum mechanical state space in the same way we do for classical mechanical state

---

[2]　Strictly speaking, the true state space is the so-called projective Hilbert space, not the full Hilbert space. In addition, the state space described here is only the space of so-called pure states. It can be argued—indeed, Chris Timpson and I argue (Wallace and Timpson 2010)—that the larger space of "mixed states" is more properly understood as the true state space. For the purposes of this chapter, though, such details just add unneeded complexity.

space. Consider (say) the state space of a single electron. There are *some* states—so-called wave packet states—such that the measurement rule predicts that the state will almost certainly have some value of position in some particular narrow range. Let ψ_x denote such a wave packet state, one for which a position measurement will almost certainly give a result close to x. Suppose, then—waving aside any qualms about "almost certainly" and "close to x"—that what ψ_x represents, physically, is a particle located at position x. Similarly, we might suppose, ψ_y represents a particle located at position y.

Here's the problem. What does $\alpha\psi_x + \beta\psi_y$ represent? By the linearity principle, it too is a possible state of the electron system. Applying the measurement algorithm to the state tells us[3] that it has probability very close to $|\alpha|^2$ of being found to have position close to x, and probability very close to $|\beta|^2$ of being found to have position close to y. So $\alpha\psi_x + \beta\psi_y$ doesn't seem to represent an electron located at any particular place in space.

Since one thing we *do* know about $\alpha\psi_x + \beta\psi_y$ is that measuring the position of a system it represents sometimes gives x and sometimes gives y, it's tempting to interpret the state space as a space of probability distributions, not as a space of actual physical states (as noted, in some other contexts in science, that's how state space works). In other words (on this approach) to say "the state of the system is $\alpha\psi_x + \beta\psi_y$," is to say something like "the electron has probability $|\alpha|^2$ of being at position x and probability $|\beta|^2$ of being at position y." (This was how Einstein hoped quantum theory would turn out; see Einstein, Podolsky, and Rosen 1935.)

This probability strategy turns out not to be viable, though, because of the quantum mechanical phenomenon called *interference*. Suppose, for instance that the dynamics of the system are such that ψ_x evolves after five seconds into $(1/\sqrt{2})$ $(\psi_x + \psi_y)$, and ψ_y evolves into $(1/\sqrt{2})(\psi_x - \psi_y)$. (This is a perfectly possible form for the dynamics of real systems.) In that case, if we delay our position measurement by five seconds, the measurement algorithm tells us that, whether the system was initially in state ψ_x or in state ψ_y, there's a 50% probability of getting a position measurement of about x, and a 50% probability of getting a position measurement of about y. ($|1/\sqrt{2}|^2 = 0.5$.)

If that's the case, and if $\alpha\psi_x + \beta\psi_y$ represents a system that is definitely in either position x or position y, we would predict that this state, too, leads to a 50% probability of each position when measured after five seconds. But in fact, if we apply linearity, we find that the *state* after five seconds is $(1/\sqrt{2})[(\alpha + \beta)\psi_x + (\alpha - \beta)\psi_y]$. If $\alpha = \beta$, for instance, the state after five seconds is just ψ_x, that is, a position measurement after five seconds will almost certainly give result x. In physics jargon, there is interference between the ψ_x and ψ_y parts of the state, so that the x outcome is *reinforced* and the y outcome is *canceled out*; interference

---

[3] Since I haven't actually stated how the measurement algorithm works, just take this on trust.

phenomena like this are very general and rule out the possibility of a probabilistic interpretation of the state space.

The only possibility left seems to be that $\alpha\psi_x + \beta\psi_y$ does represent something physical. But (1) it's not clear what—an electron in two places at once, or somehow in both places or neither, seems the only real possibility available, but it's not at all clear what that means; and (2) if electrons can be in two places at once, how is this compatible with the measurement algorithm, which says that we definitely *find* it in one place or another?

We can push this point further once we remember that measurement devices, too, are physical systems, made out of microscopic matter that is supposedly subject to the principles of quantum mechanics. As such, the process of measurement *itself* ought to be represented physically.

To see how that works, let's suppose we have a measurement device represented by a pointer that can be in three states: pointing left, pointing right, and pointing nowhere. And suppose the measurement is set up so that if the electron is measured in position x the pointer moves so that it points left, and if it is measured in position y, it moves so that it points right. We can certainly find a state space suitable for such a pointer and, indeed, can find wave packet states ϕ_L (for the pointer pointing left), ϕ_R (for it pointing right), and ϕ_0 (for it pointing nowhere). The idea of these states, as with the electron, is that ϕ_L (say) is a state such that if we measure where the pointer is—with the naked eye, or otherwise—we are pretty much guaranteed to get the result that it's in the pointing-left position.

Given state spaces for the electron and for the pointer, quantum theory gives us a recipe to construct a state space (the so-called tensor product space) for the combined system of electron plus pointer. If ϕ is any state for the electron alone, and ψ any state of the pointer alone, there is then a combined state $\phi \otimes \psi$ of both together, which gives the same experimental predictions as ϕ for measurements of the electron and the same experimental predictions as ψ for measurements of the pointer.

If the measurement device works as intended, the dynamics of measurement must look something like this:

$$\psi_x \otimes \phi_0 \rightarrow \psi_x \otimes \phi_L$$

$$\psi_y \otimes \phi_0 \rightarrow \psi_y \otimes \phi_R.$$

In other words, if the electron starts off in a state such that its position is always found to be x, the pointer must reliably end up in a state such that *its* position is always found to be on the left (and similarly for y). But now, the linearity of the dynamics causes trouble: what if we measure the electron's position when it is in the mysterious state $\alpha\psi_x + \beta\psi_y$? The dynamics in this case *have* to give

$$(\alpha\psi_x + \beta\psi_y) \otimes \phi_0 \rightarrow \alpha\psi_x \otimes \phi_L + \beta\psi_y \otimes \phi_R.$$

So there seems to be a contradiction between our measurement algorithm and the actual physical process of measurement. The algorithm tells us that the measurement should give x a fraction $|\alpha|^2$ of the time and y the rest of the time, and hence that the pointer should point left a fraction $|\alpha|^2$ of the time and right the rest of the time. But the actual physical process never gives "left" or "right" as pointer states at all and is not indeterministic at all: instead, it deterministically gives the strange, indefinite state $\alpha\psi_x \otimes \phi_L + \beta\psi_y \otimes \phi_R$, in which the pointer seems to be pointing left and pointing right at the same time.

Notice that at this stage, the strategy of interpreting the state space as a space of probabilities looks very attractive. If what it meant for the state to be $\alpha\psi_x \otimes \phi_L + \beta\psi_y \otimes \phi_R$ was that the pointer had probability $|\alpha|^2$ of pointing left and probability $|\beta|^2$ of pointing right, such states would be unmysterious. Furthermore, as a practical matter (it turns out) the interference phenomena which previously prevented us from interpreting states probabilistically do not occur at the level of macroscopic states: the dynamical process called *decoherence* (the details of which lie beyond this article[4]) guarantees that the effects of interference are invisibly weak for macroscopic systems like pointers on measurement devices.

To sum up: state spaces in nonquantum physics represent either (1) possible physical states of a system, or (2) possible probability distributions over physical states of a system. At the microscopic level, interference means that it's impossible to adopt (2), and (1) seems the only option. But at the macroscopic level, (1) seems flatly contradictory to our observations, and (2) seems a much better fit—and furthermore, the interference phenomena that rule out (2) at the microscopic level are undetectable at the macroscopic level because of decoherence.

What physicists do *in practice* is adopt (1) for microscopic systems like electrons and atoms and adopt (2) for macroscopic systems like measurement devices. The distinction between microscopic and macroscopic is rough-and-ready: basically, systems (for these purposes) count as "macroscopic" exactly when decoherence means that interference is undetectably small.

Pragmatically speaking, physics gets by fine with this dual interpretation of the state space; conceptually and philosophically speaking, though, it is profoundly unsatisfactory. Sometimes (these days, usually only in introductory textbooks) one talks about the "projection postulate" or the "collapse of the quantum state," a mysterious, indeterministic physical process that is supposed

---

[4] For those details, see Zurek (1991) or Halliwell (2010) for an introduction, Joos et al. (2003) or Schlosshauer (2007) for a systematic study, and Bacciagaluppi (2005) and Wallace (2008, pp. 22–29, 2012, chapter 3) for philosophical considerations.

to take over from the normal (linear) quantum mechanical dynamics whenever a measurement is made, and so to effect a transition from (1) to (2; but this is really just a way of applying sticking plaster to the problem, and essentially no one regards it as a principled solution.[5]

More principled solutions to the measurement problem (here I restrict my attention to solutions which avoid instrumentalism and leave classical logic alone) by and large hold on to the idea that the quantum state space represents the state of something physical, essentially because interference leaves them no choice. There are then essentially three available moves:

1. The hidden variable move: "quantum mechanics is not everything."[6] Points in quantum state space do represent the state of *something*, but they do not (at least, not alone) represent the complete state of the physical system under study. Additional information is required to do that, so the state space of quantum mechanics needs to be supplemented with a different state space, the space of states of so-called hidden variables. Normally, points in this state space encode the positions of the point particles out of which macroscopic objects are assumed to be made.[7]
2. The dynamical-collapse move: "quantum mechanics is not right." Points in quantum state space do represent the complete state of the physical system under study, but the linear dynamics are not actually correct, at least at the macroscopic level. Instead, they must be modified so that macroscopically indefinite states like $\alpha\psi_x \otimes \phi_L + \beta\psi_y \otimes \phi_R$ do not actually occur (or, if they do occur, rapidly "collapse" into definite states like $\psi_x \otimes \phi_L$).
3. The Everettian move: "quantum mechanics is everything, and it is right." Points in quantum state space do represent the complete state of the physical system under study, and the linear dynamics *are* correct. Macroscopically indefinite states like $\alpha\psi_x \otimes \phi_L + \beta\psi_y \otimes \phi_R$ are physically reasonable after all, and should be understood as describing a *multiplicity:* a situation in which there are two pointer (or two sets of pointers), one pointing left and one pointing right, and with each dynamically isolated from the other.

My concern in this chapter, of course, is exclusively with the Everettian move.

---

[5] Philosophers sometimes call this the Dirac–von Neumann interpretation of quantum mechanics. Physicists often call it the Copenhagen interpretation, but they use that term for at least two other approaches to quantum mechanics.

[6] Here and afterward I borrow from (and adapt) J. S. Bell's famous observation that "either the [quantum state together with the quantum dynamics] is not everything, or it is not right" (Bell 1987a).

[7] For technical reasons (originating, for the most part, in relativistic quantum physics), this assumption is actually pretty questionable in modern physics.

3 The Everett Interpretation

The immediate question one asks about the Everett interpretation—why do we only see one pointer, if actually there are two?—can be resolved by remembering that you, too, dear reader, are a physical system, and if χ_L and χ_R are, respectively, states in your state space representing you seeing a pointer pointing left and you seeing it pointing right, then the same linearity argument predicts that the state of (you-plus-pointer-plus-electron), once you look at the pointer, will be

$$\alpha\psi_x \otimes \phi_L \otimes \chi_L + \beta\psi_y \otimes \phi_R \otimes \chi_R.$$

In other words, *you* will be in a state of seeing left and seeing right at the same time, and this state (according to the Everett interpretation) should also be understood as telling us that there are two yous, one seeing the pointer pointing left and one seeing it pointing right.

Notice—crucially—that although the state above is the sum of two macroscopically very different state, in each term in the sum the results of the two measurements are correlated (in each term the electron has a particular position, the pointer records it as having that position, and you observe the pointer as so recording it.)

Once a system gets above a certain size, it cannot help being measured constantly—by chance collisions with the atmosphere and with sunlight, if by nothing else. In doing so, the multiplicity spreads to more and more systems, while the correlations in each term in the state remain. In due course, the state (schematically) evolves into something like

α (Whole planet is as if electron was found in position x)

$+ \beta$ (Whole planet is as if electron was found in position y).

When this, too, is understood as representing both states of affairs simultaneously, the many-worlds label for the Everett interpretation starts to sound apposite.

Of the three strategies given above, philosophers of physics—by and large—have strongly tended to prefer the first two. Physicists, on the other hand—by and large—show a general preference for the Everett interpretation (though at least as many reject the idea of a *realist* solution to the measurement problem altogether).[8] Nor is it hard to see why physicists adopt this attitude. As I have

[8] To be clear: I am not making the (relatively strong) claim that physicists prefer Everett to the various antirealist or quasi-realist strategies on the table (Copenhagen, or physics-as-information,

been at pains to stress, we specify the mathematical formalism of quantum physics by (1) providing a state space, and (2) providing a dynamics for points in that state space. Both the hidden variable and dynamic collapse strategies requires us to change one or both of these, to replace quantum theory with a new physical theory, one way or another, and physicists are generally loath to replace successful scientific theories for purely philosophical reasons, especially where—as is the case for both hidden variable and dynamical-collapse theories—no replacement is presently available.[9]

The Everett intepretation, on the other hand, is a *pure* intepretation of quantum mechanics. It leaves the quantum formalism, dynamics and state space alike, completely alone. The problems with it, if any, are entirely philosophical. From the point of view of this article, we might enumerate them as follows.

1. The problem of microscopic ontology: What are the physical properties of the possible states of a system represented by the various points of quantum mechanical state space?

2. The problem of macroscopic ontology: By virtue of what should the "wave packet" quantum states mentioned above—the states that physicists use to represent macroscopic systems with approximately determinate positions—actually be interpreted as the states of classically behaving macroscopic objects?

3. The problem of macroscopic multiplicity: By virtue of what is it the case (given (2) above) that for macroscopic systems sums of wave packet states should be interpreted as representing multiple, *non*interacting classically behaving macroscopic objects?

4. The problem of probability: By virtue of what is it that processes, like measurement, which (given (3) above) *objectively* cause the macroscopic world to split into multiple copies, are treated scientifically as probabilistic processes?

or some kind of operationalism, etc.). I am making the weaker claim that there is relatively little support for change-the-formalism strategies in mainstream physics and plenty for the Everett interpretation. You can see this in practice just by looking at the physics literature: hardly any of it explores change-the-formalism strategies, and much of it uses Everettian language in more or less explicit ways. But I base it as much on conversations with colleagues, at conferences, and so on. (It's somewhat variable by community: string theorists and quantum cosmologists are mostly Everettians; quantum information people are mostly neo-operationalists.)

[9] There are versions of both theories that are adequate for *non*relativistic physics, and there have been some encouraging signs in recent years (Dürr et al. 2004; Struyve and Westman 2007; Colin and Struyve 2007), that at least phenomenologically successful versions of hidden variable theories might be constructable in the relativistic regime. Less progress has been made for dynamic collapse theories, though Tumulka (2006) makes some interesting first steps.

Problems 2–4 are unavoidably tied up in questions of emergence, of just how facts about large-scale systems like tables and chairs (and measurement devices) supervene on facts about microphysics.[10] Here the Everett interpretation may seem to fail at the first hurdle, for there is, I think, virtually no chance of understanding macroscopic objects in quantum mechanics as some kind of mereological agglomeration of small components. However, this view of emergence is (in my view) highly unattractive even setting aside quantum mechanics, bearing as it does almost no resemblence to the way emergence actually functions in science (in, say, zoology, or fluid dynamics, or psychology, or …). There, typically, one finds that emergence is all about *structure*: theory A is emergent from theory B in some circumstance when the structural features of the world as described by theory A are instantiated in the structure of the world according to theory B. Or, at the level of objects rather than theories: higher-level objects are patterns in lower-level ones.

Granted this perspective on emergence, an answer to (2) becomes available. Wave packet states represent classical objects in motion because their dynamical behavior instantiates the dynamical behavior of those classical objects; in other words, the structural features of the classical world that justify our talk of "chairs" and "tables" and "pointers" are in fact represented—to a high degree of accuracy—by structural features of the dynamics of the wave packet states. As for (3), it follows from (again) the dynamical process of decoherence that where the quantum state of a macroscopic system is a sum of wave packets, the structural features of the macroscopic system are just the collection of all the structural features of the individually evolving wave packet states considered separately. So if a single wave packet has the right structure to instantiate a classical system, a sum of such wave packets has the right structure to instantiate a collection of independent systems. Let that "system" be something the size of a planet or solar system (or, indeed, of the universe as a whole, in principle) and that "collection of independent systems" becomes a collection of independently evolving macroscopic worlds.

This also gets us part-way toward a solution to the problem of probability. While at the fundamental level the state space and dynamics of quantum theory are those of a deterministic theory, at the emergent level, quantum mechanics has the structure of an indeterministic theory. Roughly speaking, what this means is that to every path (representing a possible history of the system) in the emergent state space, quantum theory assigns a number between zero and one, and those numbers satisfy the probability calculus: at the level of mathematical formalism (and skipping some technicalities), this is just what it *is* for a mathematical theory to be a theory of indeterministic, probabilistic dynamics.

---

[10] Here I draw heavily on my own earlier work; see Wallace 2003, 2010.

The remaining problem in (4) is again philosophical: the numbers that quantum mechanics assigns to (emergent) histories are mathematically *suitable* for representing probabilities, but *do* they represent probabilities? The question really lies beyond this chapter.[11] Suffice it to say that the contemporary literature is largely concerned with two strategies: (1) that it is justifiable for it to be a *posit* of the theory that the numbers do represent probability (or at least, that it is no less justifiable than the equivalent posit in classical mechanics); and (2) that considerations from symmetry and decision theory allow us to deduce that rational agents in a quantum universe would act *as if* the numbers represent probability. (An optional extra move is to claim that if rational agents act *as if* the numbers represent probability, that's all there is to the claim that they *do* represent it.)

None of these problems, however, address the question of the *fundamental* ontology of quantum mechanics. Furthermore, and quite apart from the intrinsic interest of the question, it might be thought that we cannot be confident in any story of emergence unless we are confident what it is emerging *from*. (Maudlin 2010, in particular, criticizes the Everett interpretation for having an inappropriate micro-ontology to appropriately ground macro-level facts; Hawthorne 2010 raises some similar concerns.) In particular, normally our concepts of space and time are treated as constant between higher-level and lower-level theories, so that for some higher-level object to exist in space-time region K it must be instantiated not just by any old objects and properties in the lower-level theory but by objects and properties themselves located in K. As such, getting some understanding of the relation between space-time and the microscopic ontology might well be crucial for the larger Everettian project. This project is my concern for the remainder of the chapter.

4 Microscopic Ontology: Problematic Strategies

Both in informal conversation and in published writings, several approaches to thinking about the micro-ontology of (Everett-interpreted) quantum mechanics frequently recur. Most of these approaches, I think, are seriously flawed, due mostly to misconceptions about quantum theory rather than internal conceptual problems. In this section I identify and criticize three such approaches.

4.1 The Eigenvalue-Eigenstate Approach

Philosophical discussions of quantum theory often refer to something called the "eigenstate-eigenvalue link," or E-E link for short.[12] In the language of this

---

[11] For extensive discussion, see Saunders et al. (2010), Greaves (2007), Wallace (2012), and references therein.

[12] For recent examples, see Albert and Loewer (1996), Barrett (1999), Lewis (2003), and Dickson (2007).

chapter, the E-E link says that a quantum state possesses a given property if and only if, according to the measurement algorithm (which, recall, associates to each quantum state a probability distribution over measurement outcomes), a measurement to detect that property would have 100% probability of so detecting it. If a system has nontrivial probability for each of several outcomes of a measurement of a given quantity, it is said, according to the E-E link, to possess an *indefinite value* of that quantity. (Monton 2006 explicitly constructs an ontology of quantum physics based on the E-E link.)

If the E-E link were correct, it would cause problems for the ontology of the Everett intepretation. In particular, no realistic macroscopic system (in the absence of wave function collapse) ever has a quantum state that would give any single macroscopic outcome 100% probability. So macroscopic systems would have indefinite values of ordinary quantities like position, and the onus would be on the Everettian to explain why we should reinterpret indefiniteness as multiplicity. Indeed, this might seem to undermine the previous section's discussion of emergent multiplicity: there aren't two pointers, there's just a pointer with an indefinite position. (Albert's "bare theory" (Albert 1992) explicitly considers the Everett interpretation from this perspective; see also chapter 4 of Barrett 1999.)

Actually, though, none of this is relevant to the Everett interpretation, because the E-E link has nothing much to do with quantum theory. One might be forgiven, if one learned one's quantum mechanics exclusively from philosophy discussions, for thinking that the E-E link was part of quantum mechanics itself, but it isn't. The measurement *algorithm* is a core part of the practice of quantum mechanics, but the E-E link is not part of that algorithm. It's an interpretive assumption. Its motivation (I think) comes from the idea that measurement must be discovering some preexisting measured value, in which case that value must be possessed by a system iff it is certain to give that value as a result of measurement. But this isn't realized in any realist interpretation of quantum mechanics—not hidden variable theories, not dynamical-collapse theories, not the Everett interpretation. It plays no part in the actual practice of physics (the term occurs just once in a search of *Physical Review* over the last century).[13] And it is anyway incompatible with the actual physics of quantities with continuously many measurement outcome possibilities, like position and momentum.[14]

For that reason, the E-E link has no particular significance for the Everett interpretation; no more, in my view, does it have any significance for quantum mechanics in general.

---

[13] In fact, the very framework used in modern physics to define the measurement algorithm—the "positive-operator-valued measure" framework—does not even allow the E-E link to be satisfactorily defined. It only makes sense if we use the somewhat outdated "projection-valued measure" framework.

[14] On a technical level, this is because no particle's wave function will remain localized in any spatial region smaller than all of space for any nonzero length of time, unless unphysical infinite potentials are used.

4.2 Reifying State Space

Some discussions of the ontology of quantum theory seem to lose track of the nature of state space. Sometimes (largely, to be fair, in conversation) one hears talk of quantum mechanics as describing the evolution of systems in Hilbert space, just as classical mechanics describes the evolution of systems in ordinary three-dimensional space. But this fails to appreciate that Hilbert space is the *state* space of quantum mechanics. To reify it—to treat it as a space of physical points, just as substantivalists do about ordinary space—is no more justified than to reify the configuration space of classical mechanics and to regard classical mechanics as about the evolution of a point in $3N$-dimensional space.[15]

Very few people are willing to defend Hilbert space realism in print. There is, however, considerable discussion of—and some support for—a more sophisticated view that, I think, suffers from a subtler form of the same fallacy. In other writing (Wallace and Timpson 2010) Chris Timpson and I have called this view *wave function realism*; we could equally call it *configuration space realism*.[16] To understand the view, it is necessary to recognize that in many (but not all) quantum systems, the quantum theory of that system is obtained by starting with a *classical* theory and then "quantizing" it. This quantizing process is mathematically at most heuristically defined and has no real conceptual justification, but it has proved highly useful as a way to construct new quantum theories. Roughly speaking, the quantum state space of a quantized theory can be represented as the space of complex functions on the configuration space of the classical theory, so that for a quantum system of N point particles, for instance, the quantized theory's state space can be represented as a space of complex functions on the $3N$-dimensional configuration space of that theory. (I say "can be represented as" rather than "is" advisedly: this is only one way to represent the structure of the quantum state space, and mathematically speaking there is no particular reason to prefer it over other representations.)

Wave function realism involves the reification of the configuration space as a physical space, so the quantum system is a complex field on this very high-dimensional space. This has led Albert (1996) to claim that according to quantum mechanics, it is an illusion that space is three-dimensional; really, it's

---

[15] An extreme proponent of structural realism might argue that it is not *wrong* to so regard classical mechanics, since the state space description is isomorphic to the more conventional description. I confess to a certain weakness for this extreme view; however, even its proponents would acknowledge that it is highly *unhelpful* to think of classical mechanics in this way. Indeed, more generally, the kind of methods that a conventional realist uses to ascertain the *true* ontology of a theory will be pretty much the same as those used by a structural realist to ascertain the most *perspicuous* way to think about that theory.

[16] Maudlin (2010) calls it *wave function monism*, though he treats it as essentially synonymous with the view that the state space of unaugmented quantum theory is the correct state space for a quantum system.

$3N$-dimensional, where N is the number of particles in the universe. This view of the ontology of (no hidden variable) quantum mechanics has probably been the most commonly assumed in the recent literature (authors who tacitly or explicitly assume it, other than Albert, have included Maudlin 2010, Hawthorne 2010, Lewis 2004, though see also Lewis's chapter in this volume, and Monton 2006).

For technical reasons (largely to do with relativistic quantum field theory; see Wallace and Timpson 2010),wave function realism is not really viable (or not without major changes) as an ontology of our universe. More important, though, it makes the same unmotivated conceptual move as Hilbert space realism: it reifies a mathematical space without any particular justification. Recall that in classical physics, configuration space is not a physical space at all: it is a space of possible classical states, and as such is very high-dimensional and very nonhomogeneous. It does not play quite that role in quantum theory; states are represented not as points in configuration space but as functions over it, and some of their structure is manifest in the structure of the function, not implicit in the structure of the space. Nonetheless, it still has a considerable role in encoding state structure: wave packet states, in particular, are featureless blobs in the configuration space representation of state space, and the structural differences between, say, a wave packet state representing a uniform gas of particles and a wave packet state representing those particles arranged into a sculpture is encoded mathematically *entirely* by the position of the respective blobs in configuration space.

As such, configuration space is not a natural candidate for reification. A strong positive case for wave function realism would be required if that position is to be taken seriously as a candidate for quantum micro-ontology. To the best of my knowledge, such a case has nowhere been presented; the position is adopted, as far as I can see, largely on the erroneous assumption that quantum theory forces it on us.

4.3   Many Worlds at the Micro Level

Multiplicity, in the Everett interpretation, is an emergent, high-level notion. The theory is a many-worlds theory in the same sense that modern astrophysics is a many-stars theory—in both cases, the objects being multiplied are not represented in the fundamental structure of the theory. Indeed, there appears to be nothing in quantum mechanics to say that a quantum state represents a universe which *fundamentally* consists of multiple classical components. (The very idea of classicality is emergent and high-level.)

However, popular accounts of the many-worlds theory—for example, Deutsch (1997)—do sometimes talk as if the multiplicity was fundamental rather than emergent. Furthermore, the configuration space representation of state space *seems* to support the idea—don't each of the points in configuration space represent a classical world?

However, the idea does not work in practice. The configuration space points themselves have no dynamics; they are simply points. The dynamics of the theory is encoded entirely in the distribution of complex numbers to the points: it is that distribution, not the points themselves, which evolves over time in accordance with the dynamical equations. Furthermore, as stressed previously, the configuration space representation is just one way to represent the quantum state space and has no special status.

This might suggest that additional dynamics, and additional rules, need to be added (the rules to pick out the configuration representation—or another representation—as preferred; the dynamics to say how each individual world evolves). At one point, the philosophical literature on the Everett interpretation was quite focused on this project (see Barrett 1999 and references therein for more details). But all such strategies abandon the idea of the Everett interpretation as a pure interpretation of quantum mechanics; instead, they become proposals for modifying the theory and, regarded as such, do not seem competitive with less ontologically expensive strategies. This kind of approach, therefore, never really found favor in physics; in philosophy, it has been moribund for some time. Contemporary defenses of the Everett interpretation, almost exclusively,[17] restrict multiplicity to the emergent level.

(I should note, however, that there is one way a *sort* of multiplicity does occur at the fundamental level in quantum theory: it occurs in the so-called sum-over-histories formalism developed by Feynman. There has been little exploration of this formalism in the philosophy literature (Arntzenius 2010 is an exception), and I have no observations to make here, except that Feynman's histories are *not,* mathematically speaking, the same as Everett's branches. The topic— though technical—might well repay further study.)

5 Microscopic Ontology: Positive Observations

I have said much in the previous section about how *not* to understand the ontology of quantum mechanics, but little positive. I finish this chapter with some provisional thoughts. To begin, recall why the Everett interpretation has a problem of microscopic ontology in the first place. The reason goes back to my original discussion of state space. Both classical and quantum theories can be formulated in state space terms; however, in the classical case the state space formalism is constructed some way into the process, once the actual metaphysical content of a "classical state" is already understood. In the quantum case, by contrast, the state space is given *first,* with even the question of *whether* points in that space represent physical states of the system being controversial.

---

[17] The only recent exception of which I am aware is Allori et al. (2011), though it is not entirely clear to me whether they actually advocate the position they present.

Once this is recognized, though, it can be seen that specifying a micro-ontology for quantum mechanics *in general* makes no more sense than doing so for classical mechanics in general. In classical mechanics, points in state space can represent (depending on the classical system in question) the length of a spring, or the positions of some particles in physical space, or the relative distances between some particles without any substantival concept of space, or the strength of some field at every point in space, or even the shape of space itself. The only things in common between these different classical mechanical theories are some rather general, abstract features of their structure and dynamics. Similarly, it is not clear to me that *anything* very substantive can be said about the ontology of quantum theory in general, over and above similarly general, similarly abstract points.

So: what *can* be said at such an abstract level? The main distinctive feature of quantum theory in general is the relation between subsystems and the overall system. In classical physics, if systems A and B have state spaces S_A and S_B, the state space of the combined system $A + B$ is just the Cartesian product of S_A and S_B: that is, the set of ordered pairs of states of A and states of B. It follows that (1) any state of the combined system uniquely specifies states of the component systems; (2) all there is to the combined system being in a given state is that its subsystems are in the appropriate states.

In quantum theory, the state space of $A + B$ is the *tensor*, rather than Cartesian, product of the separate state spaces. The mathematical details aren't crucial; what *is* crucial is that both (1) and (2) fail. This can be seen directly by looking again at the electron-plus-pointer system when its state is

$$\alpha \psi_x \otimes \phi_L + \beta \psi_y \otimes \phi_R.$$

It is fairly clear that there are no states χ, δ of electron and pointer separately such that this state can be written as $\chi \otimes \delta$. In the terminology of quantum mechanics, the two systems are *entangled*. It is really entanglement, rather than linearity and interference, that gives rise to the weirdnesses of quantum mechanics (small waves on a pond have linearity and interference!) because it is entanglement that causes microscopic indefiniteness to be magnified up to the macro level during the measurement process.

To say anything general about the metaphysics of quantum theory, then, we have to say something about the metaphysical status of entanglement. Some provisional attempts have been made; Healey (1991, 1994), for instance, regards entanglement as a form of holism and explores exactly what kind (partly through analogies with holism elsewhere in physics); see also the extensive list of references in Healey (2009). So far, however, this discussion does not seem to have made much contact with the recent ontological discussions of quantum theory in the metaphysics literature.

But to say anything that goes beyond this generality, we need to look at specific quantum theories. From the metaphysician's perspective, the natural choice is *quantum field theory,* the general framework for our current most successful quantum theories and (in particular) for the theory that underpins the Standard Model of particle physics.[18] In quantum field theory (unlike earlier versions of quantum theory), ordinary classical space-time is explicitly represented in the formalism. Indeed (speaking loosely[19]) the theory works by assigning a space of quantum states to every point of space-time. It's pretty clear what that state space represents: the state of that space-time point. But it's also necessary to remember that because of entanglement, the list of properties of each space-time point does not remotely exhaust the list of properties of the theory as a whole. Wallace and Timpson (2010) regard this as a major failure of Lewis's doctrine of Humean supervenience—the doctrine that all facts about the world supervene on monadic properties of spacetime points and the spacetime relations between them: in our view, the entanglement between (say) spacetime regions *A* and *B* should be understood precisely as encoding certain irreducible relations between *A* and *B*. Simon Saunders (1995, 1996, 1997) puts it differently (though compatibly): as he sees quantum field theory, spacetime events stand not only in spatial and temporal relations with one another but in a different kind of relation that he calls "modal" (though its relation to metaphysical modality is unclear.)

But from this point on, the devil is in the details, and it becomes impossible to discuss the topic further without delving deep into the mathematical structure of quantum theory in general and quantum field theory in particular. As this chapter has aimed to avoid such technical details, I take my leave here.

References

Albert, D. Z. (1992). *Quantum Mechanics and Experience.* Cambridge, Mass.: Harvard University Press.

Albert, D. Z. (1996). Elementary Quantum Metaphysics. In J. T. Cushing, A. Fine, and S. Goldstein (eds.), *Bohmian Mechanics and Quantum Theory: An Appraisal,* pp. 277–84. Dordrecht: Kluwer.

Albert, D. Z., and B. Loewer (1996). Tails of Schrödinger's Cat. In R. Clifton (ed.), *Perspectives on Quantum Reality,* pp. 81–92. Dordrecht: Kluwer.

Allori, V., S. Goldstein, R. Tumulka, and N. Zanghi (2011). Many-Worlds and Schrödinger's First Quantum Theory. *British Journal for the Philosophy of Science* 62:1–27.

Arntzenius, F. (2010). Quantum Mechanics, Narratability and Relativity. Unpublished manuscript

---

[18] It may seem odd that a *field* theory underpins *particle* physics. Details would take us too far afield; suffice it to say that (1) particles, too, are emergent entities in modern physics; and (2) the popular impression of particle physics as about the behavior of lots of little point particles whizzing about bears about as much relation to real particle physics as the earth/air/fire/water theory of matter bears to the Periodic Table.

[19] To be less loose would also be to court controversy; see Wallace (2011).

Bacciagaluppi, G. (2005). The Role of Decoherence in Quantum Mechanics. In *Stanford Encyclopedia of Philosophy*, available online at http://plato.stanford.edu/archives/sum2005/entries/qm-decoherence.

Barrett, J. A. (1999). *The Quantum Mechanics of Minds and Worlds*. Oxford: Oxford University Press.

Bell, J. S. (1987). Are There Quantum Jumps? In J. S. Bell, *Speakable and Unspeakable in Quantum Mechanics*, pp. 201–12. Cambridge: Cambridge University Press,.

Colin, S., and W. Struyve (2007). A Dirac Sea Pilot-Wave Model for Quantum Field Theory. *Journal of Physics A* 40:7309–42.

Deutsch, D. (1997). *The Fabric of Reality*. London: Penguin.

Dickson, M. (2007). Non-Relativistic Quantum Mechanics. In *Handbook of the Philosophy of Science: Philosophy of Physics, Part A*, pp. 275–417. Amsterdam: Elsevier.

Dürr, D., S. Goldstein, R. Tumulka, and N. Zanghí (2004). Bohmian Mechanics and Quantum Field Theory. *Physical Review Letters* 93:090402.

Einstein, A., B. Podolsky, and N. Rosen (1935). Can Quantum-Mechanical Description of Reality Be Considered Complete? *Physical Review* 47:777–80.

Greaves, H. (2007). Probability in the Everett Interpretation. *Philosophy Compass* 38–120–52.

Halliwell, J. J. (2010). Macroscopic Superpositions, Decoherent Histories and the Emergence of Hydrodynamic Behaviour. In S. Saunders, J. Barrett, A. Kent, and D. Wallace (eds.), *Many Worlds? Everett, Quantum Theory, and Reality*. Oxford: Oxford University Press.

Hawthorne, J. (2010). A Metaphysician Looks at the Everett Interpretation. In S. Saunders, J. Barrett, A. Kent, and D. Wallace (eds.), *Many Worlds? Everett, Quantum Theory, and Reality*, pp. 144–53. Oxford: Oxford University Press.

Healey, R. (1991). Holism and Nonseparability. *Journal of Philosophy* 88:393–421.

Healey, R. (1994). Nonseparable Processes and Causal Explanation. *Studies in History and Philosophy of Science* 25:337–74.

Healey, R. (2009). Holism and Nonseparability in Physics. In Edward N. Zalta (Ed.), *Stanford Encyclopedia of Philosophy (Spring 2009 edition)*, available online at http://plato.stanford.edu/archives/spr2009/entries/physics-holism.

Joos, E., H. D. Zeh, C. Kiefer, D. Giulini, J. Kupsch, and I. O. Stamatescu (2003). *Decoherence and the Appearence of a Classical World in Quantum Theory*, 2nd ed. Berlin: Springer.

Lewis, D. (1986). *Philosophical Papers, vol. II*. Oxford: Oxford University Press.

Lewis, P. J. (2003). Counting Marbles: Reply to Critics. *British Journal for the Philosophy of Science* 54:165–70.

Lewis, P. J. (2004). Life in Configuration Space. *British Journal for the Philosophy of Science* 55, 713–729.

Maudlin, T. (2010). Can the World Be Only Wavefunction? In S. Saunders, J. Barrett, A. Kent, and D. Wallace (eds.), *Many Worlds? Everett, Quantum Theory, and Reality*. Oxford: Oxford University Press.

Monton, B. (2006). Quantum Mechanics and 3N-Dimensional Space. *Philosophy of Science* 73:778–89.

Saunders, S. (1995). Time, Decoherence and Quantum Mechanics. *Synthese* 102:235–66.

Saunders, S. (1996). Time, Quantum Mechanics and Tense. *Synthese* 107:19–53.

Saunders, S. (1997). Naturalizing Metaphysics. *Monist* 80(1):44–69.

Saunders, S., J. Barrett, A. Kent, and D. Wallace (eds.) (2010). *Many Worlds? Everett, Quantum Theory, and Reality*. Oxford: Oxford University Press.

Schlosshauer, M. (2007). *Decoherence and the Quantum-to-Classical Transition*. Berlin: Springer.

Struyve, W., and H. Westman (2007). A Minimalist Pilot-Wave Model for Quantum Electrodynamics. *Proceedings of the Royal Society of London A* 463:3115–29.

Tumulka, R. (2006). Collapse and Relativity. In A. Bassi, T. Weber, and N. Zanghí (eds.), *Quantum Mechanics: Are There Quantum Jumps? and On the Present Status of Quantum Mechanics*, p. 340. American Institute of Physics Conference Proceedings 844. Available online at http://arxiv.org/abs/quant-ph/0602208.

Wallace, D. (2003). Everett and Structure. *Studies in the History and Philosophy of Modern Physics* 34:87–105.

Wallace, D. (2008). The Interpretation of Quantum Mechanics. In D. Rickles (ed.), *The Ashgate Companion to Contemporary Philosophy of Physics*, pp. 197–261. Burlington, Vt.: Ashgate.

Wallace, D. (2010). Decoherence and Ontology: Or: How I Learned to Stop Worrying and Love FAPP. In S. Saunders, J. Barrett, A. Kent, and D. Wallace (eds.), *Many Worlds? Everett, Quantum Theory, and Reality*, pp. 53–72. Oxford: Oxford University Press.

Wallace, D. (2012). *The Emergent Multiverse: Quantum Theory According to the Everett Interpretation.* Oxford: Oxford University Press.

Wallace, D. (2011). Taking Particle Physics Seriously: A Critique of the Algebraic Approach to Quantum Field Theory. *Studies in History and Philosophy of Science B* 42:116–25.

Wallace, D., and C. Timpson (2010). Quantum Mechanics on Spacetime I: Spacetime State Tealism. *British Journal for the Philosophy of Science* 61:697–727.

Zurek, W. H. (1991). Decoherence and the Transition from Quantum to Classical. *Physics Today* 43:36–44. Revised version available online at http://arxiv.org/abs/quant-ph/0306072.

Index

CPSIA information can be obtained at www.ICGtesting.com
Printed in the USA
LVOW04s1734130815

450011LV00012B/622/P